Simplify Big Data Analytics with Amazon EMR

A beginner's guide to learning and implementing
Amazon EMR for building data analytics solutions

Sakti Mishra

BIRMINGHAM—MUMBAI

Simplify Big Data Analytics with Amazon EMR

Group Product Manager: Sunith Shetty

Publishing Product Manager: Reshma Raman

Senior Editor: Tazeen Shaikh

Content Development Editor: Shreya Moharir

Technical Editor: Devanshi Ayare

Copy Editor: Safis Editing

Project Coordinator: Aparna Nair

Proofreader: Safis Editing

Indexer: Sejal Dsilva

Production Designer: Nilesh Mohite

Marketing Coordinator: Priyanka Mhatre

First published: March 2022

Production reference: 1170222

Published by Packt Publishing Ltd.

Livery Place

35 Livery Street

Birmingham

B3 2PB, UK.

ISBN 978-1-80107-107-9

www.packt.com

I dedicate this to everyone who doesn't settle down after achieving their goals but instead is encouraged to define the next one by pushing their limits.

Contributors

About the author

Sakti Mishra is an engineer, architect, author, and technology leader with over 16 years of experience in the IT industry. He is currently working as a senior data lab architect at **Amazon Web Services** (**AWS**).

He is passionate about technologies and has expertise in big data, analytics, machine learning, artificial intelligence, graph networks, web/mobile applications, and cloud technologies such as AWS and Google Cloud Platform.

Sakti has a bachelor's degree in engineering and a master's degree in business administration. He holds several certifications in Hadoop, Spark, AWS, and Google Cloud. He is also an author of multiple technology blogs, workshops, white papers and is a public speaker who represents AWS in various domains and events.

About the reviewers

Suvojit Dasgupta is a senior data architect with AWS, focusing on data engineering and analytics. In his 17 years of experience, he has led multiple strategic initiatives to design, build, migrate, modernize, and operate petabyte-scale data platforms for Fortune 500 companies. He is passionate about data architecture and takes pride in building well-architected solutions. In his free time, he likes to explore new technologies and listen to audio books. You can follow Suvojit on Twitter at @suvojitdasgupta.

Praveen Gupta is currently a data engineering manager with AWS, and has over 17 years of experience in the IT industry. Praveen started his career as an ETL/reporting developer working on traditional RDBMSs and reporting tools. Since 2014, he has been working on the AWS cloud on projects related to data science/machine learning and building complex data engineering pipelines on AWS. He specializes in data ingestion, big data processing, reporting, and building massive data warehouses at the petabyte scale for his customers, helping them make data-driven decisions. Praveen has an undergraduate degree and a master's degree, both in computer science from UIUC, USA. Praveen lives in Portland, USA with his wife and 8-year-old daughter.

Table of Contents

2

Exploring the Architecture and Deployment Options

3

Common Use Cases and Architecture Patterns

4

Big Data Applications and Notebooks Available in Amazon EMR

Section 2: Configuration, Scaling, Data Security, and Governance

5

Setting Up and Configuring EMR Clusters

6

Monitoring, Scaling, and High Availability

7

Understanding Security in Amazon EMR

8

Understanding Data Governance in Amazon EMR

Section 3: Implementing Common Use Cases and Best Practices

9

Implementing Batch ETL Pipeline with Amazon EMR and Apache Spark

10

Implementing Real-Time Streaming with Amazon EMR and Spark Streaming

14
Best Practices and Cost-Optimization Techniques

Index

Other Books You May Enjoy

Preface

As the usage of internet-related services, computers, and smart products increases, the amount of data produced by them has also increased exponentially. The data produced by them is extremely valuable for addressing business problems, as you can analyze the data to derive insights that can help in faster decision making and forecasting business growth.

These datasets are large and complex enough that traditional data processing technologies can't handle them efficiently, and that is why distributed processing frameworks such as Hadoop and Spark evolved. **Amazon Elastic MapReduce** (**EMR**) provides a managed offering for Hadoop ecosystem services, so that businesses can focus on building analytics pipelines and save time on managing infrastructure. This makes Amazon EMR the top choice for Hadoop, Spark, and big data workloads.

As the amount of data continues to grow, big data analytics will become a common skill that everybody will need to have to be successful in their career or business. Before EMR, it was expensive to try out Hadoop or Spark workloads as they require clusters of servers for setup. But with Amazon EMR's pay-as-you-go model, you can spin up small clusters quickly, scale them as needed, and terminate them when the job finishes.

Organizations that want to get started with Amazon EMR or are planning to migrate existing Hadoop workloads to EMR, as well as college-fresh graduates who want to upskill in EMR, will find this book very useful and will be able to dive deep into different EMR features and architecture patterns.

While writing this book, I have kept in mind that it should be useful to both beginners and technologists who want to learn advanced concepts of EMR. I also expect you to have some basic knowledge of AWS and Hadoop so that you can understand better and easily dive deep into advanced concepts.

By the end of this book, you will be able to comfortably architect and implement Hadoop-/Spark-based solutions with transient (job-based) or persistent (multi-tenant/long-running) EMR clusters. In addition, you will be able to understand how a complete end-to-end data analytics solution can be implemented with Amazon EMR for batch, real-time streaming, or interactive workloads. You will also gain knowledge about migration approaches, best practices, and cost optimization techniques that you can follow while implementing big data analytics workloads with EMR.

Who this book is for

This book is targeted at data engineers, data analysts, data scientists, and solution architects who are interested in building data analytics pipelines with Hadoop ecosystem services such as Hive, Spark, Presto, HBase, and Hudi. It is required that you have some prior basic knowledge of a few Hadoop ecosystem components and AWS, as well as experience with a programming language such as Python or Scala.

What this book covers

Chapter 1, An Overview of Amazon EMR, will give you an overview of Amazon EMR and its benefits compared to on-premises Hadoop clusters. Also, we will look at how EMR compares with other Spark-based AWS services such as AWS Glue and AWS Glue DataBrew.

Chapter 2, Exploring the Architecture and Deployment Options, will dive into EMR architecture; its life cycle; types of clusters; deployment options such as EMR on **Elastic Compute Cloud** (**EC2**), EMR on **Elastic Kubernetes Service** (**EKS**), and EMR on AWS Outposts; and the pricing models for each.

Chapter 3, Common Use Cases and Architecture Patterns, will explain some popular big data use cases for EMR and cover how you can build an end-to-end architecture for batch or real-time streaming and interactive analytics use cases.

Chapter 4, Big Data Applications and Notebooks Available in Amazon EMR, will give you an overview of a few of the popular Hadoop ecosystem services in EMR, such as Hive, Presto, and Spark. We will also look at a few popular machine learning frameworks, such as TensorFlow and MXNet, as well as the different notebook options available for interactive development.

Chapter 5, Setting Up and Configuring EMR Clusters, is where you will learn how to set up an EMR cluster, dive deep into its advanced configurations, learn how you can debug and troubleshoot cluster failures, and then get an overview of its SDKs and APIs.

Chapter 6, Monitoring, Scaling, and High Availability, is where you will learn about cluster monitoring, cloning, and high availability. Then, you will dive deep into different scaling aspects and understand the difference between managed and auto scaling.

Chapter 7, Understanding Security in Amazon EMR, will explain how you can make your cluster secure, covering authentication and authorization with AWS IAM, data encryption at rest and in transit, and how to leverage AWS security groups to control connectivity to your cluster.

Chapter 8, Understanding Data Governance in Amazon EMR, covers external Hive Metastore and Glue Data Catalog integration and how you can implement granular permission management with AWS Lake Formation and Apache Ranger.

Chapter 9, Implementing Batch ETL Pipeline with Amazon EMR and Apache Spark, takes you through a step-by-step guide to implementing a batch **Extract, Transform, and Load** (**ETL**) pipeline with Amazon EMR and Apache Spark.

Chapter 10, Implementing Real-Time Streaming with Amazon EMR and Spark Streaming, contains a step-by-step guide to implementing a real-time streaming ETL pipeline with Kinesis Data Stream, Amazon EMR, and Spark Streaming.

Chapter 11, Implementing UPSERT on S3 Data Lake with Apache Spark and Apache Hudi, teaches you how to do interactive development with an EMR notebook, as well as how to integrate UPSERT on an S3 data lake with Apache Hudi and Spark.

Chapter 12, Orchestrating Amazon EMR Jobs with AWS Step Functions and Apache Airflow/ MWAA, gives you an overview of AWS Step Functions, Amazon-managed Airflow, and how to integrate them to build a workflow for your EMR-based data pipelines.

Chapter 13, Migrating On-Premises Hadoop Workloads to Amazon EMR, discusses how you can migrate your on-premises Hadoop workloads to Amazon EMR, covering migrating data, catalog metadata, and ETL jobs. Also, you will learn about some of the best practices you can follow during the migration process.

Chapter 14, Best Practices and Cost Optimization Techniques, discusses some of the best practices you can follow related to EMR cluster configuration, ETL processes, file storage, and security. Then, you will learn about some of the cost optimization techniques you can follow, the limitations of EMR, and some workarounds for them.

To get the most out of this book

To follow along with the hands-on parts of the book, you need to have an AWS account with IAM permissions and an SSH client (for example, PuTTY on Windows) to connect to your EMR master node.

Software/hardware covered in the book	Operating system requirements
EMR version 6.3 to 6.5	Windows, macOS, or Linux
Spark 3.1	Windows, macOS, or Linux
Python 3/PySpark	Windows, macOS, or Linux
SSH client/PuTTY	Windows, macOS, or Linux

Before executing any of the sample code in the book, please make sure you replace the variables mentioned with your environment variables, and also make sure you have the IAM permissions required to execute the commands or scripts.

If you are using the digital version of this book, we advise you to type the code yourself or access the code from the book's GitHub repository (a link is available in the next section). Doing so will help you avoid any potential errors related to the copying and pasting of code.

The solutions given in the book are meant to give you a kick start with some sample datasets. Please move to the next stage of your learning by integrating more complex transformations that might be more applicable to your business. Also, make sure you follow the least-privileges principle while setting up production clusters.

Download the example code files

You can download the example code files for this book from GitHub at `https://github.com/PacktPublishing/Simplify-Big-Data-Analytics-with-Amazon-EMR-`. If there's an update to the code, it will be updated in the GitHub repository.

We also have other code bundles from our rich catalog of books and videos available at `https://github.com/PacktPublishing/`. Check them out!

Code in Action

The Code in Action videos for this book can be viewed at `https://bit.ly/3HM9dpj`.

Download the color images

We also provide a PDF file that has color images of the screenshots and diagrams used in this book. You can download it here: `https://static.packt-cdn.com/downloads/9781801071079_ColorImages.pdf`.

Conventions used

There are a number of text conventions used throughout this book.

`Code in text`: Indicates code words in text, database table names, folder names, filenames, file extensions, pathnames, dummy URLs, user input, and Twitter handles. Here is an example: "For example, the following sample JSON specifies configurations for the `core-site` and `mapred-site` classifications and includes Hadoop and MapReduce properties with values that you plan to override in the cluster."

A block of code is set as follows:

```
"Properties": {
  "mapred.tasktracker.map.tasks.maximum": "10",
  "mapreduce.map.sort.spill.percent": "0.80",
  "mapreduce.tasktracker.reduce.tasks.maximum": "20"
}
```

When we wish to draw your attention to a particular part of a code block, the relevant lines or items are set in bold:

```
"Classification": "core-site",
"Properties": {
  "hadoop.security.groups.cache.secs": "500"
```

Any command-line input or output is written as follows:

```
aws emr create-cluster --instance-type m5.2xlarge --release-
label emr-6.4.0 --security-configuration <mySecurityConfigName>
```

Bold: Indicates a new term, an important word, or words that you see onscreen. For instance, words in menus or dialog boxes appear in **bold**. Here is an example: "If you are creating a transient cluster that needs to execute a few steps and then auto terminate, then you can select **Step execution** for **Launch mode**."

> **Tips or Important Notes**
> Appear like this.

Get in touch

Feedback from our readers is always welcome.

General feedback: If you have questions about any aspect of this book, email us at customercare@packtpub.com and mention the book title in the subject of your message.

Errata: Although we have taken every care to ensure the accuracy of our content, mistakes do happen. If you have found a mistake in this book, we would be grateful if you would report this to us. Please visit www.packtpub.com/support/errata and fill in the form.

Piracy: If you come across any illegal copies of our works in any form on the internet, we would be grateful if you would provide us with the location address or website name. Please contact us at copyright@packt.com with a link to the material.

If you are interested in becoming an author: If there is a topic that you have expertise in and you are interested in either writing or contributing to a book, please visit authors.packtpub.com.

Share Your Thoughts

Once you've read *Simplify Big Data Analytics with Amazon EMR*, we'd love to hear your thoughts! Scan the QR code below to go straight to the Amazon review page for this book and share your feedback.

https://packt.link/r/1-801-07107-1

Your review is important to us and the tech community and will help us make sure we're delivering excellent quality content.

Section 1: Overview, Architecture, Big Data Applications, and Common Use Cases of Amazon EMR

This section will provide an overview of Amazon EMR, along with its architecture, cluster nodes, features, benefits, different deployment options, and pricing. Then it will provide an overview of different big data applications EMR supports and showcase common architecture patterns we see with Amazon EMR.

This section comprises the following chapters:

- *Chapter 1, An Overview of Amazon EMR*
- *Chapter 2, Exploring the Architecture and Deployment Options*
- *Chapter 3, Common Use Cases and Architecture Patterns*
- *Chapter 4, Big Data Applications and Notebooks available in Amazon EMR*

1
An Overview of Amazon EMR

This chapter will provide an overview of Amazon **Elastic MapReduce** (**EMR**), its benefits related to big data processing, and how its cluster is designed compared to on-premises Hadoop clusters. It will then explain how Amazon EMR integrates with other **Amazon Web Services** (**AWS**) services and how you can build a Lake House architecture in AWS.

You will then learn the difference between the Amazon EMR, **AWS Glue**, and **AWS Glue DataBrew** services. Understanding the difference will make you aware of the options available when deploying **Hadoop** or **Spark** workloads in AWS.

Before going into this chapter, it is assumed that you are familiar with Hadoop-based big data processing workloads, have had exposure to AWS basis concepts, and are looking to get an overview of the Amazon EMR service so that you can use it for your big data processing workloads.

The following topics will be covered in this chapter:

- What is Amazon EMR?
- Overview of Amazon EMR
- Decoupling compute and storage
- Integration with other AWS services

- EMR release history

- Comparing Amazon EMR with AWS Glue and AWS Glue DataBrew

What is Amazon EMR?

Amazon EMR is an AWS service that provides a distributed cluster for big data processing. Now, before diving deep into EMR, let's first understand what big data represents, for which EMR is a solution or tool.

What is big data?

The beginnings of enormous volumes of datasets date back to the 1970s, when the world of data was just getting started with data centers and the development of relational databases, despite the fact that the concept of **big data** was still relatively new. These technology revolutions led to personal desktop computers, followed by laptops, and then mobile computers over the next several decades. As people got access to devices, the data being generated started growing exponentially.

Around the year 2005, people started to realize that users generate huge amounts of data. Social platforms, such as Facebook, Twitter, and YouTube generate data faster than ever, as users get access to smart products or internet-related services.

Put simply, big data refers to large, complex datasets, particularly those derived from new data sources. These datasets are large enough that traditional data processing software can't handle its storage and processing efficiently. But these massive volumes of data are of great use when we need to derive insights by analyzing them and then address business problems with it, which we were not able to do before. For example, an organization can analyze their users' or customers' interactions with their social pages or website to identify their sentiment against their business and products.

Often, big data is described by the five Vs. It started with three Vs, which includes data *volume*, *velocity*, and *variety*, but as it evolved, the accuracy and value of data also became major aspects of big data, which is when *veracity* and *value* got added to represent it as five Vs. These five Vs are explained as follows:

- **Volume**: This represents the amount of data you have for analysis and it really varies from organization to organization. It can range from terabytes to petabytes in scale.

- **Velocity**: This represents the speed at which data is being collected or processed for analysis. This can be a daily data feed you receive from your vendor or a real-time streaming use case, where you receive data every second to every minute.

- **Variety**: When we talk about variety, it means what the different forms or types of data you receive are for processing or analysis. In general, they are broadly categorized into the following three:

  - **Structured**: Organized data format with a fixed schema. It can be from relational databases or CSVs or delimited files.

  - **Semi-structured**: Partially organized data that does not have a fixed schema, for example, XML or JSON files.

  - **Unstructured**: These datasets are more represented through media files, where they don't have a schema to follow, for example, audio or video files.

- **Veracity**: This represents how reliable or truthful your data is. When you plan to analyze big data and derive insights out of it, the accuracy or quality of the data matters.

- **Value**: This is often referred to as the worth of the data you have collected as it is meant to give insights that can help the business drive growth.

With the evolution of big data, the primary challenge became how to process such huge volumes of data, because the typical single system processing frameworks were not enough to handle them. It needed a distributed processing computing framework that can do parallel processing.

After understanding what big data represents, let's look at how the Hadoop processing framework helped to solve this big data processing problem statement and why it became so popular.

Hadoop – processing framework to handle big data

Though there were different technologies or frameworks that came to handle big data, the framework that got the most traction is **Hadoop**, which is an open source framework designed specifically for storing and analyzing big datasets. It allows combining multiple computers to form a cluster that can do parallel distributed processing to handle gigabyte-to petabyte-scale data.

The following is a data flow model that explains how the input data is collected, stored into **Hadoop Distributed File System** (**HDFS**), then processed with Hive, Pig, or Spark big data processing frameworks and the transformed output becomes available for consumption or is transferred to downstream systems or external vendors. It represents a high-level data flow, where input data is collected and stored as raw data. It then gets processed as needed for analysis and then made available for consumption:

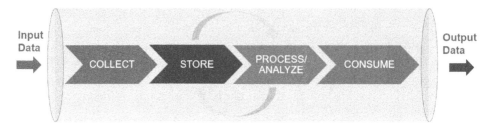

Figure 1.1 – Data flow in a Hadoop cluster

The following are the main basic components of Hadoop:

- **HDFS**: A distributed filesystem that runs on commodity hardware and provides improved data throughput as compared to traditional filesystems and higher reliability with an in-built fault tolerance mechanism.

- **Yet Another Resource Negotiator** (**YARN**): When multiple compute nodes are involved with parallel processing capability, YARN helps to manage and monitor compute CPU and memory resources and also helps in scheduling jobs and tasks.

- **MapReduce**: This is a distributed framework that has two basic modules, that is, *map* and *reduce*. The *map* task reads the data from HDFS or a distributed storage layer and converts it into key-value pairs, which then becomes input to the *reduce* tasks, which ideally aggregates the map output to provide the result.

- **Hadoop Common**: These include common Java libraries that can be used across all modules of the Hadoop framework.

In recent years, the Hadoop framework became popular because of its **massively parallel processing** (**MPP**) capability on top of commodity hardware and its fault-tolerant nature, which made it more reliable. It was extended with additional tools and applications to form an ecosystem that can help to collect, store, process, analyze, and manage big data. Some of the most popular applications are as follows:

- **Spark**: An open source distributed processing system that uses in-memory caching and optimized execution for fast performance. Similar to MapReduce, it provides batch processing capability as well as real-time streaming, machine learning, and graph processing capabilities.

- **Hive**: Allows users to use distributed processing engines such as MapReduce, Tez, or Spark to query data from the distributed filesystem through the SQL interface.

- **Presto**: Similar to Hive, Presto is also an open source distributed SQL query engine that is optimized for low-latency data access from the distributed filesystem. It's used for complex queries, aggregations, joins, and window functions. The Presto engine is available as two separate components in EMR, that is, PrestoDB and PrestoSQL or Trino.

- **HBase**: An open source non-relational or NoSQL database that runs on top of the distributed filesystem that provides fast lookup for tables with billions of rows and millions of columns grouped as column families.

- **Oozie**: Enables workflow orchestration with Oozie scheduler and coordinator components.

- **ZooKeeper**: Helps in managing and coordinating Hadoop component resources with inter-component-based communication, grouping, and maintenance.

- **Zeppelin**: An interactive notebook that enables interactive data exploration using Python and PySpark kind of frameworks.

Hadoop provides a great solution to big data processing needs and it has become popular with data engineers, data analysts, and data scientists for different analytical workloads. With its growing usage, Hadoop clusters have brought in high maintenance overhead, which includes keeping the cluster up to date with the latest software releases and adding or removing nodes to meet the variable workload needs.

Now let's understand the challenges on-premises Hadoop clusters face and how Amazon EMR comes as a solution to them.

Challenges with on-premises Hadoop clusters

Before Amazon EMR, customers used to have on-premises Hadoop clusters and faced the following issues:

- **Tightly coupled compute and storage architecture**: Clusters used to use HDFS as their storage layer, where the data node's disk storage contributes to HDFS. In the case of node failures or replacements, there used to be data movement to have another replica of data created.

- **Overutilized during peak hours and underutilized at other times**: As the autoscaling capabilities were not there, customers used to do capacity planning beforehand and add nodes to the cluster before usage. This way, clusters used to have a constant number of nodes; during peak usage hours, cluster resources were overutilized and during off-hours, they were underutilized.

- **Centralized resource with the thrashing of resources**: As resources get overutilized during peak hours, this leads to the thrashing of resources and affects the performance or collapse of hardware resources.

- **Difficulty in upgrading the entire stack**: Setting up and configuring services was a tedious task as you needed to install specific versions of Hadoop applications and when you planned to upgrade, there were no options to roll back or downgrade.

- **Difficulty in managing many different deployments (dev/test)**: As the cluster setup and configuration was a tedious task, developers didn't have the option to quickly build applications in new versions to prove feasibility. Also, spinning up different development and test environments was a time-consuming process.

To overcome the preceding challenges, AWS came up with Amazon EMR, which is a managed Hadoop cluster that can scale up and down as workload resource needs change.

Overview of Amazon EMR – managed and scalable Hadoop cluster in AWS

To give an overview, Amazon EMR is an AWS tool for big data processing that provides a managed, scalable Hadoop cluster with multiple deployment options that includes EMR on Amazon **Elastic Compute Cloud** (**EC2**), EMR on Amazon **Elastic Kubernetes Service** (**EKS**), and EMR on AWS Outposts.

Amazon EMR makes it simple to set up, run, and scale your big data environments by automating time-consuming tasks such as provisioning instances, configuring them with Hadoop services, and tuning the cluster parameters for better performance.

Amazon EMR is used in a variety of applications, including **Extract, Transform, and Load** (**ETL**), clickstream analysis, real-time streaming, interactive analytics, machine learning, scientific simulation, and bioinformatics. You can run petabyte-scale analytics workloads on EMR for less than half the cost of traditional on-premises solutions and more than three times faster than open source Apache Spark. Every year, customers launch millions of EMR clusters for their batch or streaming use cases.

Before diving into the benefits of EMR compared to an on-premises Hadoop cluster, let's look at a brief history of Hadoop and EMR releases.

A brief history of the major big data releases

Before we go further, the following diagram shows the release period of some of the major databases:

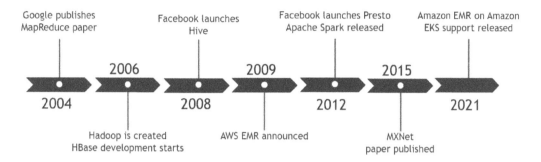

Figure 1.2 – Diagram explaining the history of major big data releases

As you can see in the preceding diagram, Hadoop was created in 2006 based on Google's MapReduce whitepaper and then AWS launched Amazon EMR in 2009. Since then, EMR has added a lot of features and its recent launch of Amazon EMR on EKS provides the great capability to run Spark workloads in Kubernetes clusters.

Now is a good time to understand the benefits of Amazon EMR and how its cluster is configured to decouple compute and storage.

Benefits of Amazon EMR

There are numerous advantages of using Amazon EMR, and this section provides an overview of these advantages. This will in turn help you when looking for solutions based on Hadoop or Spark workloads:

- **Easy to use**: You can set up an Amazon EMR cluster in minutes without having to worry about provisioning cluster instances, setting up Hadoop configurations, or tuning the cluster.

  You get the ability to create an EMR cluster through the AWS console's **user interface (UI)**, where you have both quick and advanced options to specify your cluster configurations, or you can use AWS **command-line interface (CLI)** commands or AWS SDK APIs to automate the creation process.

- **Low cost**: Amazon EMR pricing is based on the infrastructure on top of which it is deployed. You can choose from the different deployment options EMR provides, but the most popular usage pattern is with Amazon EC2 instances.

When we configure or deploy a cluster on top of Amazon EC2 instances, the pricing depends on the type of EC2 instance and the Region you have selected to launch your cluster. With EC2, you can choose on-demand instances or you can reduce the cost by purchasing reserved instances with a commitment of usage. You can lower the cost even further by using a combination of spot instances, specifically while scaling the cluster with task nodes.

- **Scalability**: One of the biggest advantages of EMR compared to on-premises Hadoop clusters is its elastic nature, using which you can increase or decrease the number of instances of your cluster. You can create your cluster with a minimal number of instances and then can scale your cluster as the job demands. EMR provides two scalability options, autoscaling and managed scaling, which scales the cluster based on resource utilization.

- **Flexibility**: Though EMR provides a quick cluster creation option, you have full control over your cluster and jobs, where you can make customizations in terms of setup or configurations. While launching the cluster, you can select the default Linux **Amazon Machine Images (AMIs)** for your instances or integrate custom AMIs and then install additional third-party libraries or configure startup scripts/ jobs for the cluster.

  You can also use EMR to reconfigure apps on clusters that are already running, without relaunching the clusters.

- **Reliability**: Reliability is something that is built into EMR's core implementation. The health of cluster instances is constantly monitored by EMR and it automatically replaces failed or poorly performing instances. Then new tasks get instantiated in newly added instances.

  EMR also provides multi-master configuration (up to three master nodes), which makes the master node fault-tolerant. EMR also keeps the service up to date by including stable releases of the open source Hadoop and related application software at regular intervals, which reduces the maintenance effort of the environment.

- **Security**: EMR automatically configures a few default settings to make the environment secure, including launching the cluster in Amazon **Virtual Private Cloud (VPC)** with required network access controls and configuring security groups for EC2 instances.

  It also provides additional security configurations that you can utilize to improve the security of the environment, which includes enabling encryption through AWS KMS keys or your own managed keys, configuring strong authentication with Kerberos, and securing the in-transit data through SSL.

You can also use **AWS Lake Formation** or **Apache Ranger** to configure fine-grained access control on the cluster databases, tables, or columns. We will dive deep into each of these concepts in later chapters of the book.

- **Ease of integration**: When you build a data analytics pipeline, apart from EMR's big data processing capability, you might also need integration with other services to build the production-scale implementation.

  EMR has native integration with a lot of additional services and some of the major ones include orchestrating the pipeline with **AWS Step Functions** or Amazon **Managed Workflows for Apache Airflow** (**MWAA**), close integration with AWS IAM to integrate tighter security control, fine-grained access control with AWS Lake Formation, or developing, visualizing, and debugging data engineering and data science applications built in R, Python, Scala, and PySpark using the EMR Studio **integrated development environment** (**IDE**).

- **Monitoring**: EMR provides in-depth monitoring and audit capability on the cluster using AWS services such as CloudWatch and CloudTrail.

  CloudWatch provides a centralized logging platform to track the performance of your jobs and cluster and define alarms based on specific thresholds of specific metrics. CloudTrail provides audit capability on cluster actions. Amazon EMR also has the ability to archive log files in Amazon **Simple Storage Service** (**S3**), so you can refer to them for debugging even after your cluster is terminated.

Apart from CloudWatch and CloudTrail, you can also use the Ganglia monitoring tool to monitor cluster instance health, which is available as an optional software configuration when you launch your cluster.

Decoupling compute and storage

When you integrate an EMR cluster for your batch or streaming workloads, you have the option to use the core node's HDFS as your primary distributed storage or Amazon S3 as your distributed storage layer. As you know, Amazon S3 provides a highly durable and scalable storage solution and Amazon EMR natively integrates with it.

With Amazon S3 as the cluster's distributed storage, you can decouple compute and storage, which gives additional flexibility. It enables you to integrate job-based transient clusters, where S3 acts as a permanent store and the cluster core node's HDFS is used for temporary storage. This way, you can decouple different jobs to have their own cluster with the required amount of resources and scaling in place and avoid having an always-on cluster to save costs.

The following diagram represents how multiple transient EMR clusters that contain various steps can use S3 as their common persistent storage layer. This can also help for disaster recovery implementation:

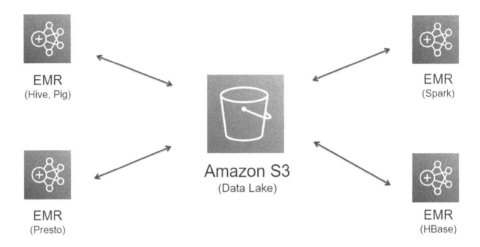

Figure 1.3 – Multiple EMR clusters using Amazon S3 as their distributed storage

Now that you understand how EMR provides flexibility to decouple compute and storage, in the next section, you will learn how you can use this feature to create persistent or transient clusters depending on your use case.

Persistent versus transient clusters

Persistent clusters represent a cluster that is always active to support multi-tenant workloads or interactive analytics. These clusters can have a constant node capacity or a minimal set of nodes with autoscaling capabilities. Autoscaling is a feature of EMR, where EMR automatically scales up (adds nodes) or scales down (removes nodes) cluster resources based on a few cluster utilization parameters. In future chapters, we will dive deep into EMR scaling features and options.

Transient clusters are treated more as job-based clusters, which are short-lived. They get created with data arrival or through scheduled events, do the data processing, write the output back to target storage, and then get terminated. These also have a constant set of nodes to start with and then scale to support the additional workloads. But when you have transient cluster workloads, ideally Amazon S3 is used as a persistent data store so that after cluster termination, you still have access to the data to perform additional ETL or business intelligence reporting.

Here is a diagram that represents different kinds of cluster use cases you may have:

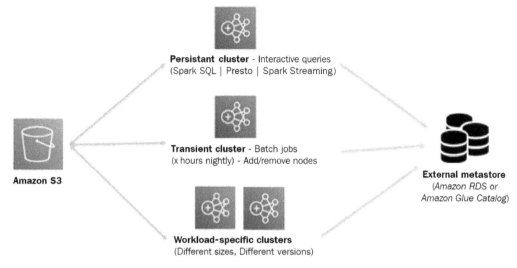

Figure 1.4 – EMR architecture representing cluster nodes

As you can see, all three clusters are using Amazon S3 as their persistent storage layer, which decouples compute and storage. This will facilitate scaling for both compute and storage independently, where Amazon S3 provides scaling with 99.999999999% (11 9s) durability and the cluster compute capacity can scale horizontally by adding more core or task nodes.

As represented in the diagram, transient clusters can be scheduled jobs or multiple workload-specific clusters running in parallel to do ETL on their datasets, where each workload cluster might have workload-specific cluster capacity.

When you implement transient clusters, often the best practice is to externalize your Hive Metastore, which means if your cluster gets terminated and becomes active again, it does not need to create Metastore or catalog tables again. When you are externalizing Hive Metastore of your EMR cluster, you have the option to use an Amazon RDS database as a Hive Metastore or you can use AWS Glue Data Catalog as your Metastore.

Integration with other AWS services

By now, you have got a good overview of Amazon EMR and its architecture, which can help you visualize how you can execute your Hadoop workloads on Amazon EMR.

But when you build an enterprise architecture for a data analytics pipeline, be it batch or real-time streaming, there are a lot of additional benefits to running in AWS. You can decouple your architecture into multiple components and integrate various other AWS services to build a fault-tolerant, scalable architecture that is highly secure.

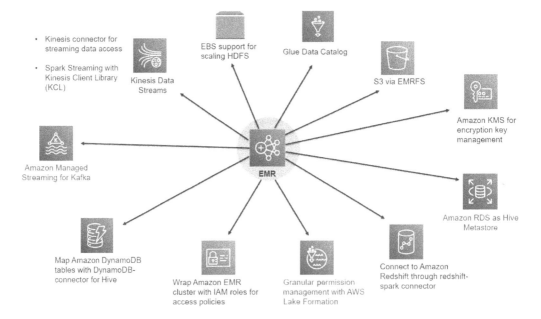

Figure 1.5 – Representing EMR integration with other AWS services

The preceding figure is a high-level diagram that shows how you can integrate a few other AWS services with Amazon EMR for an analytics pipeline. These are just a few sets of services listed to give you an idea, but there are a lot of other AWS services that you can integrate which you deem fit for your use case.

Now let's get an overview of these services and understand how they integrate with Amazon EMR.

Amazon S3 with EMR File System (EMRFS)

Out of all the AWS services, Amazon S3 takes the top spot as any data analytics architecture built on top of AWS will have S3 as a persistent or intermediate data store. When we build a data processing pipeline with Amazon EMR, S3 integration is natively supported through **EMR File System** (**EMRFS**). When a job communicates with an Amazon S3 path to read or write data, it can access S3 with the s3 :// prefix.

Amazon Kinesis Data Streams (KDS)

Amazon **Kinesis Data Streams** (**KDS**) is a commonly used messaging service within AWS to build real-time streaming pipelines for use cases such as website clickstreams, application log streams, and **Internet of Things** (**IoT**) device event streams. It is scalable and durable and continuously captures gigabytes of data per second with multiple sources ingesting to it and multiple consumers reading from it in parallel.

It provides **Kinesis Producer Library** (**KPL**), which data producers can integrate to push data to Kinesis, and also provides **Kinesis Consumer Library** (**KCL**), which data-consuming applications can integrate to access the data.

When we build a real-time streaming pipeline with EMR and KDS as a source, we can use Spark Structured Streaming, which integrates KCL internally to access the stream datasets.

Amazon Managed Streaming for Kafka (MSK)

Similar to KDS, **Apache Kafka** is also a popular messaging service in the open source world that is capable of handling massive volumes of data for real-time streaming. But it comes with the additional overhead of managing the infrastructure.

Amazon **Managed Streaming for Kafka** (**MSK**) is a fully managed service built on top of open source Apache Kafka that automates Kafka cluster creation and maintenance. You can set up a Kafka cluster with a few clicks and use that as an event message source when you plan to implement a real-time streaming use case with EMR and Spark Streaming as the processing framework.

AWS Glue Data Catalog

AWS Glue is a fully managed ETL service that is built on top of Apache Spark with additional functionalities, such as Glue crawlers and Glue Data Catalog. **Glue crawlers** help autodetect the schema of source datasets and create virtual tables in **Glue Data Catalog**.

With EMR 5.8.0 or later, you can configure Spark SQL in EMR to use AWS Glue Data Catalog as its external metastore. This is great when you have transient cluster scenarios that need an external persistent metastore or multiple clusters sharing a common catalog.

Amazon Relational Database Service (RDS)

Similar to Glue Data Catalog, you can also use Amazon **Relational Database Service (RDS)** to be the external metastore for Hive, which can be shared between multiple clusters as a persistent metastore.

Apart from being used as an external metastore, in a few use cases, Amazon RDS is also used as an operational data store for reporting to which data gets ingested through EMR big data processing, which pushes aggregated output to RDS for real-time reporting.

Amazon DynamoDB

Amazon DynamoDB is an AWS-hosted, fully managed, scalable NoSQL database that delivers quick, predictable performance. As it's serverless, it takes away the infrastructure management overhead and also provides all security features, including encryption at rest.

In a few analytical use cases, DynamoDB is used to store data ingestion or extraction-related checkpoint information and you can use DynamoDB APIs with Spark to query the information or define Hive external tables with a DynamoDB connector to query them.

Amazon Redshift

Amazon Redshift is an MPP data warehousing service of AWS using which you can query and process exabytes of structured or semi-structured data. In the data analytics world, having a data warehouse or data mart is very common and Redshift can be used for both.

In the data analytics use cases, it's a common pattern that after your ETL pipeline processing is done, the aggregated output gets stored in a data warehouse or data mart and that is where the EMR-to-Redshift connection comes into the picture. Once EMR writes output to Redshift, you can integrate business intelligence reporting tools on top of it.

AWS Lake Formation

AWS Lake Formation is a service that enables you to integrate granular permission management on your data lake in AWS. When you define AWS Glue Data Catalog tables on top of a data lake, you can use AWS Lake Formation to define access permissions on databases, tables, and columns available in the same or other AWS accounts. This helps in having centralized data governance, which manages permissions for AWS accounts across an organization.

In EMR, when you try to pull data from Glue Data Catalog tables and use it as an external metastore, then your EMR cluster processes such as Spark will go through Lake Formation permissions to access the data.

AWS Identity and Access Management (IAM)

AWS **Identity and Access Management** (**IAM**) enables you to integrate authentication and authorization for accessing AWS services through the console or AWS APIs. You can create groups, users, or roles and define policies to give or restrict access to specific resources or APIs.

While creating an EMR cluster or accessing its API resources, every request goes through IAM policies to validate the access.

AWS Key Management Service (KMS)

When you think of securing your data while it's being transferred through the network or being stored in a storage layer, you can think of cryptographic keys and integrating an encryption and decryption mechanism. To implement this, you need to store your keys in a secured place that integrates with your application well and AWS **Key Management Service** (**KMS**) makes that simple for you. AWS KMS is a highly secure and resilient solution that protects your keys with hardware security modules.

Your EMR cluster can interact with AWS KMS to get the keys for encrypting or decrypting the data while it's being stored or transferred between cluster nodes.

Lake House architecture overview

Lake House is a new architecture pattern that tries to address the shortcomings of data lakes and combines the best of data lakes and data warehousing. It acknowledges that the one-size-fits-all strategy to analytics eventually leads to compromises. It is not just about connecting a data lake to a data warehouse or making data lake access more structured; it's also about connecting a data lake, a data warehouse, and other purpose-built data storage to enable unified data management and governance.

In AWS, you can use Amazon S3 as a data lake, Amazon EMR or AWS Glue for ETL transformations, and Redshift for data warehousing. Then, you can integrate other relational NoSQL data stores on top of it to solve different big data or machine learning use cases.

The following diagram is a high-level representation of how you can integrate the Lake House architecture in AWS:

Figure 1.6 – Lake House architecture reference

As you can see in the preceding diagram, we have the Amazon S3 data lake in the center, supported by AWS Glue for serverless ETL and AWS Lake Formation for granular permission management.

Around the centralized data lake, we have the following:

- Amazon EMR for batch or streaming big data processing

- Amazon OpenSearch service for log analytics or search use cases

- Amazon Redshift for data warehousing or data mart use cases

- Amazon DynamoDB for key-value NoSQL store

- Amazon Aurora for operational reporting or external metastore

- Amazon SageMaker for machine learning model training and inference

As explained previously, the Lake House architecture represents how you can bring in the best of multiple services to build an ecosystem that addresses your organization's analytics needs.

EMR release history

As Amazon EMR is built on top of the open source Hadoop ecosystem, it tries to stay up to date with the open source stable releases, which includes new features and bug fixes.

Each EMR release comprises different Hadoop ecosystem applications or services that fit together with specific versions. EMR uses **Apache Bigtop**, which is an open source project within the Apache community to package the Hadoop ecosystem applications or components for an EMR release.

When you launch a cluster, you need to select the EMR cluster version and with advanced options, you can identify which version of each Hadoop application is integrated into that EMR release. If you are using AWS SDK or AWS CLI commands to create a cluster, you can specify the version using the *release label*. Release labels follow a naming convention of `emr-x.x.x`, for example, `emr-6.3.0`.

The EMR documentation clearly lists each release version and the Hadoop components integrated into it.

The following is a diagram of the EMR 6.3.0 release, which lists a few components of Hadoop services that are integrated into it and how it compares to previous releases of EMR 6.x:

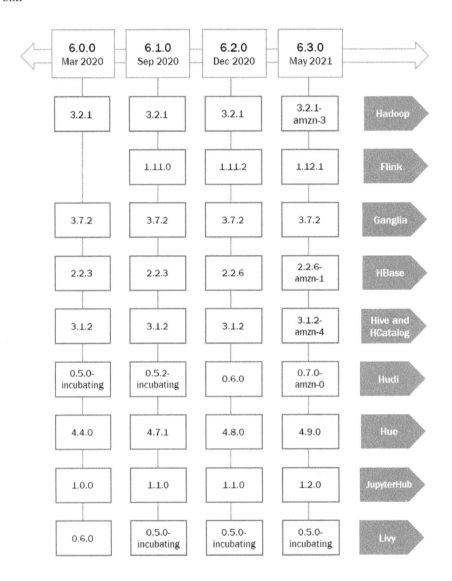

Figure 1.7 – Diagram of EMR release version comparison

If you were using open source Hadoop or any third-party Hadoop clusters and then migrating to EMR, it is best to go through the release documentation, understand different versions of Hadoop applications integrated into it, find the different configurations involved related to security, network access, authentication, authorization, and so on, and then evaluate it against your current Hadoop cluster to plan for migration.

With this, you have got a good overview of Amazon EMR, its benefits, its release history, and more. Now, let's compare it with a few other AWS services that are also based on Spark workloads and understand how they compare with Amazon EMR.

Comparing Amazon EMR with AWS Glue and AWS Glue DataBrew

When you look at today's big data processing frameworks, Spark is very popular for its in-memory processing capability. This is because it gives better performance compared to earlier Hadoop frameworks, such as MapReduce.

Earlier, we talked about different kinds of big data workloads you might have; it could be batch or streaming or a persistent/transient ETL use case.

Now, when you look for AWS services for your Spark workloads, EMR is not the only option AWS provides. You can use AWS Glue or AWS Glue DataBrew as an alternate service too. Customers often get confused between these services, and knowing what capabilities each of them has and when to use them can be tricky.

So, let's get an overview of these alternate services and then talk about what features they have and how to choose them by use case.

AWS Glue

AWS Glue is a serverless data integration service that is simple to use and is based on the Apache Spark engine. It enables you to discover, analyze, and transform the data through Spark-based in-memory processing. You can use AWS Glue for exploring datasets, doing ETL transformations, running real-time streaming pipelines, or preparing data for machine learning.

AWS Glue has the following components that you can benefit from:

- **Glue crawlers and Glue Data Catalog**: AWS Glue crawlers provide the benefit of deriving a schema from an S3 object store, where they scan a subset of data and create a table in Glue Data Catalog, on top of which you can execute SQL queries through Amazon Athena.

- **Glue Studio and jobs**: Glue Studio provides a visual interface to design ETL pipelines, which autogenerates PySpark or Scala scripts, which you can modify to integrate your complex business logic for data integration.

- **Glue workflows**: This enables you to build workflow orchestration for your ETL pipeline that can integrate Glue crawlers or jobs to be executed in sequence or parallel.

Please note, AWS Glue is a serverless offering, which means you don't have access to the underlying infrastructure and its pricing is based on **Data Processing Units** (**DPUs**). Each unit of DPU comprises 4 vCPU cores and 16 GB memory.

Example architecture for a batch ETL pipeline

Here is a simple reference architecture that you can follow to build a batch ETL pipeline. The use case is when data lands into the Amazon S3 landing zone from different sources and you need to build a centralized data lake on top of which you plan to do data analysis or reporting:

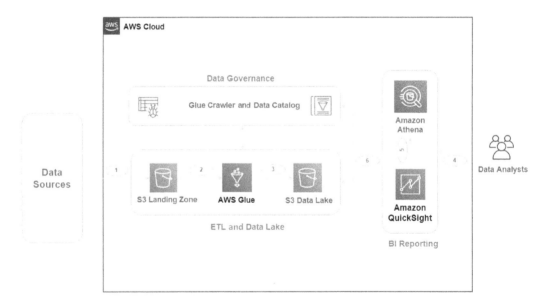

Figure 1.8 – Example architecture representing an AWS Glue ETL pipeline

As you can see in the diagram, we have the following steps:

- **Step 1**: Data lands into the Amazon S3 landing zone from different data sources, which becomes the raw zone for the data.

- **Step 2-3**: You will be using Glue crawlers and jobs to apply data cleansing, standardization, and transformation, and then make it available in an S3 data lake bucket for consumption.

- **Step 4-6**: Integrates flow to consume the data lake data for data analysis and business reporting. As you can see, we have integrated Amazon Athena to query data from Glue Data Catalog and S3 through standard SQL and integrated Amazon QuickSight for business intelligence reporting.

If you note, Glue crawlers and Glue Data Catalog are represented as a common centralized component for ETL transformations and data analysis. As your storage layer is Amazon S3, defining virtual schema on top of it will help you to access data through SQL as you do in relational databases.

AWS Glue DataBrew

AWS Glue DataBrew is a visual data preparation tool that assists data analysts and data scientists prepare data for data analysis or machine learning model training and inference. Often, data scientists spend 80% of their time preparing the data for analysis and 20% of the time on model development.

AWS Glue DataBrew solves that problem, where data scientists can save the effort of the steps from custom coding to clean, normalized data by building a transformation rule on the visual UI in minutes. AWS Glue DataBrew has 250+ prebuilt transformations (for example, filtering, adding derived columns, filtering anomalies, correcting invalid values, and joining or merging different datasets) that you can use to clean or transform your data, and it converts the visual transformation steps into a Spark script under the hood, which gives you faster performance.

It saves the transformation rules as recipes that you can apply to multiple jobs and can configure your job output format, partitioning strategy, and execution schedule. It also provides additional data profiling and lineage capability.

AWS Glue DataBrew is serverless, so you don't need to worry about setting up a cluster or managing its infrastructure resources. Its pricing is pretty similar to other AWS services, where you only pay for what you use.

Example architecture for machine learning data preparation

The following is a simple reference architecture that represents a data preparation use case for machine learning prediction and inference:

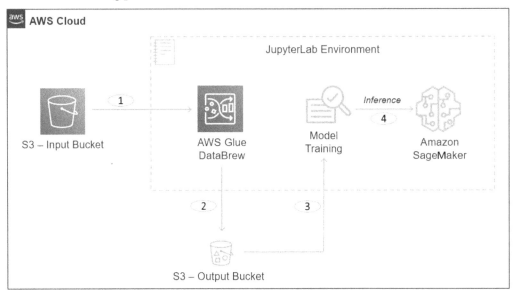

Figure 1.9 – An overall architecture representing data preparation with AWS Glue DataBrew

As you can see in the diagram, we have the following steps:

- **Steps 1-2**: Represents AWS Glue DataBrew reading data from the S3 input bucket and, after processing, writing the output back to the S3 output bucket

- **Steps 3-4**: Represents Amazon SageMaker using the processed data of the data bucket for machine learning training and inference, which also integrates Jupyter Notebook for model development

Now, let's look at how to decide which service is best for your use case.

Choosing the right service for your use case

Now, after getting an overview of all three AWS services, you can take note of the following guidelines when choosing the right service for your use case:

- **AWS Glue DataBrew**: If you are trying to build an ETL job or pipeline with Spark but you are new to Hadoop/Spark or you are not good at writing scripts for ETL transformations, then you can go for AWS Glue Data Brew, where you can use the GUI-based actions to preview your data and apply necessary transformation rules.

This is great when you receive different types of file formats from different systems and don't want to spend time writing code to prepare the data for analysis:

- **Pros**: Does not require you to learn Spark or scripting languages for preparing your data and also, you can build a data pipeline faster.

- **Cons**: Just because you are relying on the UI actions to build your pipeline, you lose the flexibility of building complex ETL operations that are not available through the UI. Also, it does not support real-time streaming use cases.

- **Target users**: Data scientists or data analysts can take advantage of this service as they spend time preparing the data or cleansing it for analysis and their objective is not to apply complex ETL operations.

- **Use cases**: Data cleansing and preparation with minimal ETL transformations.

- **AWS Glue**: If your objective is to build complex Spark-based ETL transformations by joining different data sources and you are looking for a serverless solution to avoid infrastructure management hassles, then AWS Glue is great.

  On top of the Spark-based ETL job capability, AWS Glues crawlers, Glue Data Catalog, and workflows are also great benefits:

  - **Pros**: Great for serverless Spark workloads that support both batch and streaming pipelines. You can use AWS Glue Studio to generate base code, on top of which you can edit.

  - **Cons**: AWS Glue is limited to only Spark workloads and with Spark, you can use only Scala and Python. Also, if you have persistent cluster requirements, Glue is not a great choice.

  - **Target users**: Data engineers looking for Spark-based ETL engines are best suited to use AWS Glue.

  - **Use cases**: Batch and streaming ETL transformations and building a unified data catalog.

- **Amazon EMR**: As you have understood by now, AWS Glue or AWS Glue DataBrew are great for Spark-based workloads only and are great if you are looking for serverless options. But there are a lot of other use cases where organizations go with a combination of different Hadoop ecosystem services (for example, Hive, Flink, Presto, HBase, TensorFlow, and MXNet) or would like to have better control of not only the infrastructure, instance type and so on but also specific versions of Hadoop/Spark services they would like to use.

Also, sometimes you will have use cases where you might look for persistent Hadoop clusters that need to be used by multiple teams for different purposes, such as data analysis/preparation, ETL, real-time streaming, and machine learning models. EMR is a great fit there:

- **Pros**: Gives control to choose cluster capacity, instance types, and Hadoop services you need with version selection and also provides auto- and managed scaling features. Also provides flexibility to use spot instances for cost savings and have better control of the network and security of your cluster.

- **Cons**: Not a serverless offering like AWS Glue, but that's the purpose of EMR, to give you better control to configure your cluster.

- **Target users**: EMR can be used by mostly all kinds of users who deal with data on a daily basis, such as data engineers, data analysts, and data scientists.

- **Use cases**: Batch and real-time streaming, machine learning, interactive analytics, genomics, and so on.

I hope this gave you a good understanding of how you can choose the right AWS service for your Hadoop or Spark workloads and also how they compare with each other in terms of features, pros, and cons.

Summary

Over the course of this chapter, we got an overview of the Hadoop ecosystem and EMR and learned about its benefits and the problem statement it solves.

After covering those topics, we got an overview of other AWS services that integrate with EMR to build an end-to-end AWS cloud-native architecture. We followed that with a discussion of the Lake House architecture and EMR releases.

Finally, we covered how EMR compares with other Spark-based AWS services, such as AWS Glue and AWS Glue DataBrew, and how to choose the right service for your use case.

That concludes this chapter! Hopefully, you have got a good overview of Amazon EMR and are now ready to dive deep into its architecture and deployment options, which will be covered in the next chapter.

Test your knowledge

Before moving on to the next chapter, test your knowledge with the following questions:

1. You have an on-premises persistent Hadoop cluster, where you have a lot of Hive SQL jobs and very few Spark ETL jobs are available. This cluster serves multiple teams and also helps in interactive analytics.

   You are assigned the job to plan for AWS cloud migration. Which AWS service is best suited for you?

2. You have received a JSON file from your source system that you would like to flatten and apply a few standardizations to for your machine learning model prediction. This process needs to be repeated every day so that the machine learning prediction can predict the output for the next day. Which AWS service will you use?

3. You have a requirement to build a real-time streaming application, where you need to integrate a scalable message bus and Spark Streaming consumer application. You are looking for a managed messaging service that can scale as the number of streaming events increase or decrease. What will you use?

4. Which AWS services will you choose for the message bus and Spark processing consumer application?

Further reading

The following are a few resources you can refer to for further reading:

- About EMR releases: `https://docs.aws.amazon.com/emr/latest/ReleaseGuide/emr-release-components.html`

- EMR release history: `https://docs.aws.amazon.com/emr/latest/ReleaseGuide/emr-whatsnew-history.html`

2
Exploring the Architecture and Deployment Options

This chapter will dive deep into the **Elastic MapReduce** (**EMR**) architecture. We will also look at the different deployment options it provides, such as Amazon EMR on **Amazon Elastic Compute Cloud** (**EC2**), Amazon EMR on **Amazon Elastic Kubernetes Service** (**EKS**), and Amazon EMR on **AWS Outposts**. It will also explain details around different EMR cluster node types, its life cycle, and ways to submit work to the cluster.

Toward the end of the chapter, you will learn how EMR pricing works with different deployment options and how you can use **AWS Budgets** and **Cost Explorer** for cost-related monitoring.

As we proceed to further chapters of this book, where we will cover different use cases and implementation patterns around EMR, an understanding of the architecture and deployment options will be a prerequisite.

The following topics will be covered in this chapter:

- EMR architecture deep dive
- Understanding clusters and nodes
- Using S3 versus HDFS for cluster storage
- Understanding the cluster life cycle
- Building Hadoop jobs with dependencies in a specific EMR release version
- EMR deployment options

> **Important Note**
> It is assumed that you are familiar with the Hadoop ecosystem architecture and this chapter will primarily focus on architecture changes with Amazon EMR.

EMR architecture deep dive

The following is a high-level architecture of Amazon EMR, which includes various components, such as the distributed storage layer, cluster resource management with **Yet Another Resource Negotiator** (**YARN**), batch or stream processing frameworks, and different Hadoop applications.

Apart from these major components, the following architecture also represents monitoring with Ganglia, the Hue user interface, Zeppelin notebook, Livy server, and connectors that enable integration with other AWS services:

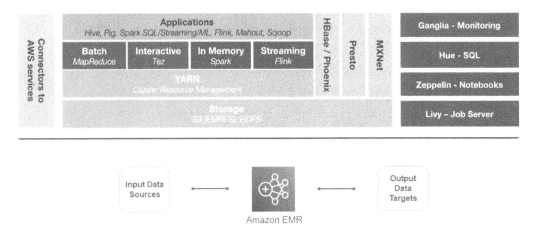

Figure 2.1 – EMR high-level architecture representing core components and applications

Now let's discuss each of these components in detail.

Distributed storage layer

In a typical on-premises Hadoop cluster or Hadoop on EC2 architectures, you will notice the Hadoop cluster node's disk space contributes to **Hadoop Distributed File System (HDFS)** storage space, and the storage and compute are tightly coupled.

But EMR also provides the option to decouple your compute from storage. Now, let's look at each storage option and understand which use cases they can be a fit for.

HDFS as cluster storage

HDFS is a distributed filesystem of Hadoop with horizontal scaling capabilities, which means as you add more nodes to the cluster you get additional compute and storage capacity.

HDFS distributes the data across cluster nodes as blocks, which are on average between 64 MB and 128 MB in size, and also enables you to have multiple copies of data that are fault-tolerant. The default configuration is to maintain three copies of data distributed into different nodes, but you can increase or decrease it based on your use case.

HDFS is great as *persistent storage* when you have higher random read writes or you have defined **Service-Level Agreements (SLAs)** around accessing the data. As the data is stored local to the instance, you get better performance while accessing it.

S3 with EMR File System (EMRFS) as cluster storage

EMR File System (EMRFS) is an extended filesystem of Hadoop created by Amazon to integrate Amazon **Simple Storage Service (S3)** as the permanent storage for EMR. This integration is seamless as you just need to use EMRFS with the s3a:// or s3n:// or s3:// S3 prefix in the cluster and all your cluster jobs will start pointing to S3.

When you use S3 with EMRFS, HDFS is still being used as intermediate storage during job execution and the final output is being written to EMRFS.

A node's local filesystem

When you use Amazon EC2 instances, it comes with a preattached disk that is called an instance store, and then you can attach additional ephemeral disk volume, which is called **Elastic Block Store (EBS)**. When you talk about HDFS as file storage, it generally refers to the EBS volumes attached to instances.

But you can also refer to the instance store volume for your HDFS, but that is generally not recommended as the data in it is retained only during the life cycle of the instance. As soon as you terminate or restart the instance, you lose the data in it.

YARN – cluster resource manager

In Hadoop clusters, YARN is one of the major components as it helps to manage the cluster resources and also coordinates job execution across multiple nodes.

YARN became very popular because of its multi-tenancy feature, which allows execution of batch, streaming, and graph processing jobs, its optimized resource management, and its scaling capability.

Some of the major components of YARN are an **ApplicationMaster**, a **NodeManager**, a **ResourceManager**, and **containers**. These containers include disk, memory, and CPU resources of a node and the ResourceManager is used to coordinate all the resources required for different job execution. The ApplicationMaster works with the NodeManager for job execution, its monitoring and completion, and it gets required resources from the ResourceManager.

In Amazon EMR, by default, most of the Hadoop applications or frameworks use YARN, but there are a few others that don't use YARN to handle their resources. On each node, EMR runs an agent that manages YARN components and communicates with Amazon EMR.

If you recollect, we explained in the previous chapter how you can use **EC2 spot instances** for EMR task nodes to save more costs and make it more scalable. But with spot instances, the chances of task failure are high as they get terminated because of resource unavailability. Now, to make the jobs fault-tolerant, EMR allows running the ApplicationMaster in core nodes only, so that spot node termination will not terminate the ApplicationMaster and it can trigger the failed job in another node.

With the 5.19.0 release, EMR introduced a built-in YARN node label feature using which it labels core nodes with the CORE label and configures yarn-site, and capacity-schedulers to make use of these labels and make sure the ApplicationMaster runs only these nodes.

> **Important Note**
>
> Manually overriding or modifying the yarn-site or capacity-scheduler configuration files of the cluster that have CORE node labels integrated into it might break the feature, which allows running the ApplicationMaster only in core nodes.

Distributed processing frameworks

While designing big data analytics applications, depending on the use case, you might look for different batch and real-time streaming frameworks, and EMR provides a few options around it. A few of the frameworks use YARN and a few others use their own ResourceManager. Depending on the framework you integrate for your data processing, you will have programming language options as not all frameworks support all languages.

Out of the different frameworks, MapReduce and Spark are very common and these days, Spark is widely used for most batch and real-time streaming use cases.

MapReduce

Hadoop **MapReduce** is one of the popular open source frameworks that has **map** and **reduce** as two primary steps. In the *map* step, it reads input data as per the block size defined in the Hadoop configuration files and output key values pairs. The *reduce* step takes the map step output as input, does the defined aggregations, and then writes the output as part files to HDFS or S3. The number of reducers defines the number of output files you will have and with configuration parameters, you can control how many reducers you need for your job.

Hadoop applications such as Hive and Pig use the MapReduce framework as their processing engine to do transformations.

Spark

Similar to the Hadoop MapReduce framework, **Spark** is also another open source framework that is widely used for big data processing. Spark became more popular compared to MapReduce because of its directed acyclic graph execution, faster in-memory processing, support for different programming languages, such as Java, Scala, Python, and R, and multiple APIs to support batch and real-time streaming and graph processing kinds of use cases.

When you use Spark in EMR, you have native integration with EMRFS to read from and write data to S3.

Hadoop applications

Amazon EMR supports many Hadoop ecosystem applications to serve data collection, processing, analysis, or consumption needs. Each of these applications has its own API interface and programming language support.

A few of the popular applications are MapReduce, Tez, and Spark, which are used for big data processing. Sqoop is used for pulling data from relational databases, TensorFlow and MXNet are used for machine learning, Spark Streaming and Flink are used for real-time streaming, Hive and Presto are used as query engines, and HBase is used as a NoSQL database on HDFS or S3.

We will dive deep into a few of these applications in the upcoming chapters.

With this, you should have a good understanding of EMR's overall architecture and an understanding of each of its components. To understand more about the EMR cluster, next we will dive deep into its cluster node types and how they are structured.

Understanding clusters and nodes

The primary construct or component of Amazon EMR is the **cluster**, and the cluster is a collection of Amazon EC2 instances, which are called **nodes**. Each node within the cluster has a type, depending on the role it plays or the job it does in the cluster. Based on the node type, respective Hadoop libraries are installed and configured on that instance.

The following are the node types available in EMR:

- **Master node**: Master nodes are responsible for managing cluster instances, monitoring health, coordinating job execution, tracking the status of tasks, and so on. This is a must-have node type when you create a cluster and you can have a single node cluster with just a master node in it.

- **Core node**: This node type is responsible for storing data in the HDFS on your cluster and runs Hadoop application services such as Hive, Pig, HBase, and Hue. If you have a multi-node cluster, then you should have at least one core node.

- **Task node**: This node type is responsible for executing tasks with the amount of CPU or memory it has. Task nodes are optional and are useful when you plan to increase your cluster capacity for a specific job and scale down after its completion. These node types do not have HDFS storage.

The following diagram represents the master node, core node, and task node of a cluster with both HDFS and S3 as the storage layer options:

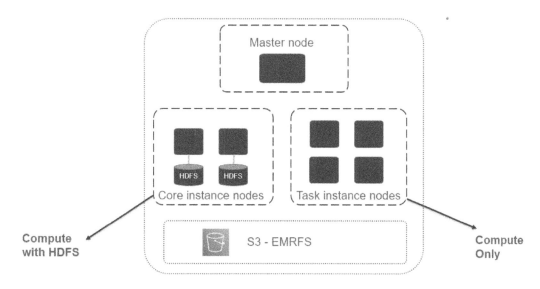

Figure 2.2 – EMR architecture representing cluster nodes

While creating an EMR cluster, you have two configuration options that you can apply to each of the cluster node types (master/core/task). Either you can select a uniform type of instance, which is called an **instance group**, or you can select a mix of different instance types, which is called an **instance fleet**. The configuration you select will be applied for the duration of the cluster and an instance fleet and instance group cannot coexist in your cluster.

Uniform instance groups

When creating an EMR cluster, you have the flexibility to group different instance types and assign core or task node roles to them. This way, you are not restricted to selecting one instance type for your whole cluster.

In general practice, you can select different EC2 instance types for the master node, core nodes, and task nodes. This also helps when you plan to integrate autoscaling into your cluster and you can scale your task nodes, which will be using instances that have higher compute and memory capacity and less disk capacity as they won't have HDFS. The following diagram shows EMR cluster nodes with instance groups:

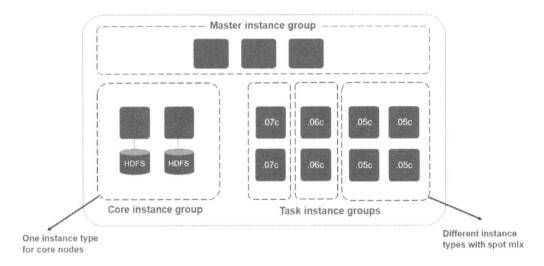

Figure 2.3 – EMR cluster nodes with instance groups

As you can see in the diagram, each node type is grouped as an instance group, and a task instance group has multiple instance types combined to form subgroups. Core instance groups have both compute and HDFS storage, whereas task instance groups have only compute capacity with the option to choose spot instances.

Instance fleet

With instance fleet configuration, you can combine up to five instance types in a single fleet and assign to them a node type. Using a task instance fleet is optional but it provides flexibility to create a mix with spot instance types too. Each instance can have a different EBS volume configuration.

You can define the target capacity for on-demand and spot instances in terms of vCPU cores, and EMR will select any combination of specified instance types to meet the target capacity. For master nodes, specify a single instance type so that it is consistent, and it would be better if you go for the on-demand instance type only.

In this section, we have learned about the EMR cluster's node types and how they are configured using instance groups or instance fleets. Next, we will get an overview of cluster storage, where you can use HDFS cluster storage or Amazon S3.

Using S3 versus HDFS for cluster storage

As you may have understood by now, EMR has the flexibility to choose HDFS or EMRFS + S3 as the cluster's persistent storage. As explained previously, EMR has different types of nodes: the master node, core nodes, and task nodes.

Now, let's understand how both of these storage layers are different and which problem statements they solve.

HDFS as cluster-persistent storage

As you can see from the following diagram, there are multiple core nodes pointing to the master node, and each core node has its own CPU, memory, and HDFS storage:

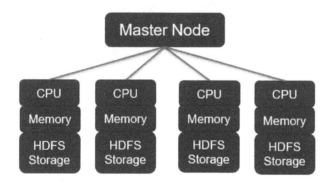

Figure 2.4 – EMR node structure with HDFS as persistent storage

These are some properties to be aware of when your cluster uses HDFS as persistent storage:

- You need to maintain by default three copies of data across the core nodes to be fault-tolerant.

- An EMR cluster is deployed in a single **Availability Zone** (**AZ**) of a Region, so a complete AZ failure might cause data loss.

- As HDFS is formed with the core nodes' EBS volumes, your storage cost will depend on the EBS volumes.

- Data is stored locally, which means the cluster needs to be available 24x7 even if no jobs are running, utilizing the cluster capacity.

Now, let's look at Amazon S3 as a storage layer.

Amazon S3 as a persistent data store

The following architecture diagram represents the integration of Amazon S3 as the persistent data store instead of HDFS, where all core nodes or task nodes will interact with the S3 prefix to read or write data:

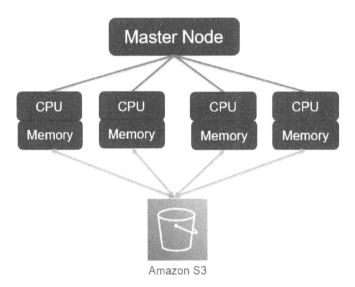

Figure 2.5 – EMR node structure with Amazon S3 as persistent storage

The following are a few of the benefits you get when you use Amazon S3 as the persistent storage layer:

- With S3 being a persistent store, you get more reliability because of S3 multi-AZ replication.

- Your cluster is protected against data loss from node failure, cluster failure, or AZ failure.

- Compared to EBS-based HDFS, S3 is much cheaper, which brings your total costs down.

- As compute and storage are decoupled, you can terminate clusters when idle or multiple clusters can point to the same dataset in S3.

In this section, we have explained what cluster storage options you have and what benefits or tradeoffs they have while integrating it. In the next section, we will dive deep into the EMR cluster's life cycle and how you can submit jobs to the cluster as steps.

> **Important Note**
>
> Amazon S3 has a limit around the maximum number of write or read requests you can get per second. It is 3,500 for PUT/COPY/POST/DELETE and 5,500 for GET/HEAD requests per second per prefix in a bucket. To avoid hitting the maximum limit, you can think of adding more S3 prefixes while writing output, you can think of reducing the number of write or read requests per second, or you can also think of increasing the EMRFS retry limit.

Understanding the cluster life cycle

When you launch an EMR cluster through the AWS SDK, **command-line interface (CLI)**, or console, it follows a series of steps to launch required infrastructure resources, configure them with required libraries, and then execute any bootstrap actions defined.

The following is the sequence of steps the cluster follows to complete the setup successfully:

1. Provision the EC2 instances for the cluster to represent master, core, and task nodes using the default AMI or the custom API you have specified. At this phase, the cluster shows the status as **STARTING**.

2. Run the bootstrap actions that you have specified to install custom third-party libraries or do additional configurations on instances, or start any specific services. At this phase, the cluster shows the status as **BOOTSTRAPPING**.

3. Install libraries related to the Hadoop services (Hive, Pig, Hue, Spark, HBase, Tez, and so on) you have selected during the cluster launch. After completion of this step, the cluster state is **WAITING** if no jobs are submitted for execution.

After the cluster is ready and in the **WAITING** state, you can submit jobs to the cluster through the AWS CLI, SDK, or console, and each job is treated as a step. It can be a Hive, Pig, or Spark step that reads from HDFS or S3, does the **Extract, Transform, and Load (ETL)** operation, and writes the data back to the storage layer. Following the completion of the step, again the cluster goes back to the **WAITING** state.

While creating the cluster, you can set it to auto-terminate once the last step is performed, which is better suited for transient job-based cluster use cases. When the cluster gets a termination request, its state goes to **TERMINATING**, and then after successful termination, it goes to **TERMINATED**.

During the launch, if the cluster creation fails because of any error, then Amazon EMR terminates the cluster, and the state of the cluster is set to **TERMINATED_WITH_ERRORS**. Please note, you do have the option to enable **Termination Protection**, which means in case of failures, the cluster will not get terminated, and in such scenarios, you can manually disable termination protection on the cluster, then trigger the termination action.

The following diagram represents the life cycle of a cluster, which means the sequence of steps EMR takes to set up the cluster, configure it, and execute jobs, and what the cluster state is during each stage:

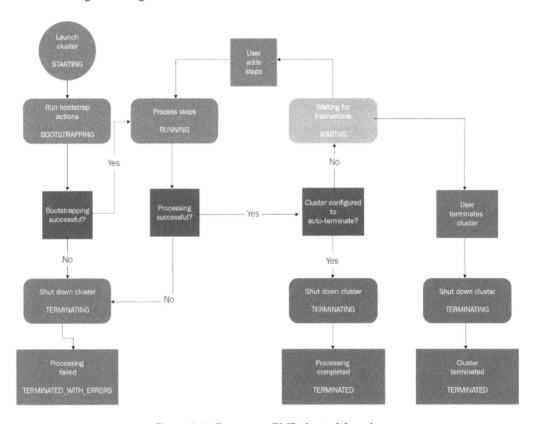

Figure 2.6 – Represents EMR cluster life cycle

Now, as we understand the cluster life cycle and the sequence of steps it takes for setup, next we will learn how you can submit jobs to the cluster and what the steps a job execution goes through are.

Options to submit work to the cluster

You have several options to submit work to the EMR cluster. A few of them are listed here:

- When you have a persistent EMR cluster, you can submit jobs through the AWS console or AWS CLI commands, or submit dynamically from applications using EMR APIs.

- In the case of a persistent cluster, you can also SSH to the master node or the respective Hadoop application's core node and use the CLI of the Hadoop application to submit queries or jobs.

- If you have a transient EMR cluster, then you can include the job triggering steps as part of the cluster creation script or command, which will create the cluster, submit the job as a step, then, post-completion, terminate the cluster.

- You can also invoke cluster creation or job submission actions from workflow orchestration tools such as **AWS Step Functions** and self-managed or AWS-managed **Apache Airflow** clusters. For persistent clusters, orchestration tools trigger job submission commands, and for transient cluster use cases, you can trigger cluster creation, followed by job submission, and then termination.

Next, let's understand the steps of job execution.

Submitting jobs to the cluster as EMR steps

When you design an ETL pipeline with multiple transformation jobs, you can submit each job as a step to the cluster and each job can invoke different Hadoop services.

For example, you can have the following two steps in your cluster, which flattens a nested JSON file to derive some insights:

- A **PySpark** job that reads nested JSON from S3 and flattens it out as a fixed schema file, then writes the output back to S3

- A **Hive** job that defines an external table on top of the step-1 output S3 path and does SQL-based aggregations to create summarized output and, finally, writes the output back to S3

Ideally, before triggering the EMR steps, you will upload the nested JSON file to the input S3 bucket, which the PySpark jobs will read, and also create the intermediate and final output S3 buckets or paths that the Spark and Hive step will use.

This way, we are decoupling the compute and storage by using S3 as the permanent storage layer.

The following is the sequence EMR takes to run a step:

1. Request submitted to start the processing steps.

2. The state of both the PySpark and Hive steps is set to **PENDING**.

3. When the first PySpark step goes into execution, its state gets changed to **RUNNING**.

4. After the PySpark state completes, its state changes to **COMPLETED**, and the Hive step's state changes to **RUNNING** as that's defined as the next step.

5. When the Hive job execution completes, its state changes to **COMPLETED** too.

6. This pattern gets repeated for every step until they are all marked as **COMPLETED**, and finally, the cluster gets terminated if auto-terminate is set to **TRUE**.

The following diagram represents the sequence of states each step goes through when it is getting processed:

Figure 2.7 – Sequence diagram for Amazon EMR showing the different cluster step states

As you can see from the preceding diagram, **Step 1** starts processing and then moves to the **COMPLETED** state. That triggers the execution of **Step 2**, which is in the **RUNNING** state, and the rest of the states are in the **PENDING** state, waiting for **Step 2** to complete. In ETL pipelines, failures are pretty common and can be because of resource unavailability, data corruption, or schema mismatch issues. You do have the option to specify what will happen if a particular step fails, which will be marked as the **FAILED** status. You can either choose to ignore the failure and proceed with the next steps or mark the rest of the remaining steps as **CANCELLED** and proceed with cluster termination. In the case of failures, the default behavior is to mark the remaining steps as **CANCELLED**.

The following diagram represents the step sequence when a particular step fails processing:

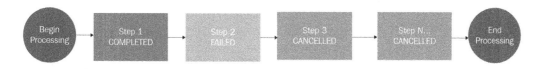

Figure 2.8 – Sequence diagram for Amazon EMR showing failed step

As you can see in this diagram, **Step 2** is marked as **FAILED** and after that, all other steps are **CANCELLED** and that marks the end of the processing.

This section described the EMR cluster life cycle flow and how job submission works with the status of each job or step. Next, you will learn how you should include job-dependent libraries while building Hadoop jobs in specific EMR releases.

Building Hadoop jobs with dependencies in a specific EMR release version

When you build different Hadoop, Hive, or Spark jobs and execute them on a specific version of the EMR cluster, you might often face version conflict issues between your application code and its dependencies because the specific versions of libraries your code expects might not be available in the cluster. So, it's necessary that you build your application code against the libraries available in the cluster.

Starting with the Amazon EMR 5.18.0 release, you can integrate the *Amazon EMR artifact repository*, using which you can build your application to avoid version conflicts or runtime classpath errors when you execute them in the EMR cluster.

You can add the artifact repository to your Maven project or with pom.xml, which has the following syntax:

```
https://<s3-endpoint>/<region-ID-emr-artifacts>/<emr-
release-label>/repos/maven/
```

Now, let's understand each parameter of the preceding https URL, which will help you form your repository URL:

- s3-endpoint is the Amazon S3 endpoint that follows its Region-specific URL format, for example, s3.us-east-1.amazonaws.com for us-east-1 Regions. Because there are no differences in artifacts between Regions, you can choose the one that best suits your environment.

- `Emr-release-label` is the Amazon EMR cluster's release label and, as highlighted in the previous chapter, release labels follow a format of `emr-x.x.x`, for example, `emr-6.3.0`. A specific EMR release series may include multiple release versions, but you can use the first release version within that series. For example, use `emr-5.30.0` for the EMR cluster EMR5.30.1 version.

As an example, if you are using the EMR release version 5.30.1 in the `us-east-1` Region, then your URL will be as follows:

```
https://s3.us-east-1.amazonaws.com/us-east-1-emr-artifacts/emr-
5.30.0/repos/maven/
```

Next, we can look at the different deployment options EMR provides and how their pricing is calculated.

EMR deployment options

As Amazon EMR is built on top of the open source Hadoop ecosystem, it tries to stay up to date with the open source stable releases, which includes new features and bug fixes.

Amazon EMR on Amazon EC2

Amazon EMR on Amazon EC2 is the first deployment option EMR offered and is very popular across different use cases. With EC2, you get the broadest range of instance types, which you can select depending on your workload and use case to get the best performance and cost benefits.

The following is a sample AWS CLI command that creates an Amazon EMR cluster with the `emr-6.3.0` release label, five `m5.xlarge` instances, and a Spark application:

```
$ aws emr create-cluster \
--name "First EMR on EC2 Cluster" \
--release-label emr-6.3.0 \
--applications Name=Spark \
--ec2-attributes KeyName=<myKeyPairName> \
--instance-type m5.xlarge \
--instance-count 5 \
--use-default-roles
```

Before executing the preceding command, please replace the `<myKeyPairName>` variable with your EC2 key pair name.

We will go deeper into the EMR on EC2 deployment option and its configuration later in the book.

EC2 instance types to support different workloads

EMR provides flexibility to select a variety of EC2 instance families for different workloads. A few are listed here:

- **General-purpose**: For typical batch ETL pipelines, you can select from the M4 and M5 EC2 instance families, which are geared toward general batch processing.

- **Compute-intensive**: For compute-intensive workloads, for example, machine learning jobs, you can use the C4 or C5 instance types.

- **Memory-intensive**: For high memory usage applications, such as Spark-based heavy ETL workloads or interactive low-latency query requirements, you can use the R4 or X1 instance families.

- **Dense disk storage needs**: For workloads that need higher storage capacity for HDFS, you can select from the D2 or I3 instance families, which come with higher EBS storage.

> Important Note
>
> The EC2 instance families listed here are based on the availability while writing this book and they are subject to change as EMR starts supporting new instance types.

Now let's look at another deployment option.

Amazon EMR on Amazon EKS

EMR on EKS provides great value. It helps if you already have an Amazon EKS cluster that is running different workloads on other applications and you would like to use the same cluster for Spark workloads. With EMR on EKS, you can automate the provisioning of Spark workloads and also use the Amazon EMR optimized runtime for Apache Spark to accelerate your workloads by up to three times.

With EMR on EKS, you can achieve multiple other benefits:

- You can save time from managing open source Spark on EKS, and can on developing an application with Spark.

- You can choose any specific EMR + Spark version with EKS, which gives you an EMR Spark runtime that is three times faster than open source Spark.

- You can use the same EKS cluster with isolation for different Spark workloads and can have control over granular access permissions.

Let's now look at the architecture in this case.

The architecture of an EMR on EKS cluster

The following is a high-level diagram that explains how you can submit different data engineering jobs to the EMR virtual cluster backed by EKS. The EKS cluster can be configured to run with EC2 instances or AWS Fargate and you can choose different Spark versions for different applications:

Figure 2.9 – High-level architecture diagram representing the EMR on EKS deployment option

The following are the high-level components of an EMR on EKS cluster:

- **Kubernetes namespace**: Amazon EKS uses Kubernetes namespaces to create isolation between applications or users within the cluster. While deploying an Amazon EKS cluster, you have the option to select Amazon EC2 or AWS Fargate as its backend compute layer, which you can specify as a Kubernetes namespace.

- **Virtual cluster**: The EMR on EKS cluster you create is called a virtual cluster as it does not create any resources and uses a Kubernetes namespace with which it is registered. You can have multiple virtual clusters created pointing to the same EKS cluster with their own namespace.

- **Job run**: This represents submitting a job to the EMR virtual cluster, which submits it to the backend EKS cluster. This job can be a Spark job. At the time of writing this book, EMR on EKS supports Spark only and we can only hope for additional Hadoop application support in the future.

- **Managed endpoints**: For interactive analytics, you can use a managed endpoint that integrates with EMR Studio and submits the job execution to the underlying EKS cluster.

Next, let's look at an example.

Example AWS CLI commands to manage the cluster and jobs

Assuming you already have an EMR on EKS cluster, you should use the following AWS CLI commands to interact with the EMR virtual cluster.

The following are three sample AWS CLI commands to create an EMR on EKS cluster, submit a job, and then terminate the cluster, to represent a transient EMR cluster use case:

1. Create a virtual cluster with the EKS namespace:

```
$ aws emr-containers create-virtual-cluster \
--name <virtual_cluster_name> \
--container-provider '{
  "id": "<eks_cluster_name>",
  "type": "EKS",
  "info": {
    "eksInfo": {
      "namespace": "<namespace_name>"
    }
```

```
        }
    }'
```

2. Trigger a PySpark job:

```
$ aws emr-containers start-job-run \
--name <job_name> \
--virtual-cluster-id <cluster_id> \
--execution-role-arn <IAM_role_arn> \
--release-label <emr_release_label> \
--job-driver '{
  "sparkSubmitJobDriver": {
    "entryPoint": <entry_point_location>,
    "entryPointArguments": ["<arguments_list>"],
    "sparkSubmitParameters": <spark_parameters>
  }
}' \
--configuration-overrides '{
  "monitoringConfiguration": {
    "cloudWatchMonitoringConfiguration": {
      "logGroupName": "<log_group_name>",
      "logStreamNamePrefix": "<log_stream_prefix>"
    }
  }
}'
```

3. Delete the EMR virtual cluster:

```
aws emr-containers delete-virtual-cluster —id <cluster_
id>
```

After every job run, you can delete the EMR virtual cluster, but it's recommended to create the cluster once and keep it active for multiple job runs. As the EMR virtual cluster consumes no resources and does not add to the cost, keeping it active will reduce the overhead of creating and deleting it multiple times.

There are a few additional commands you can use to list, monitor, or cancel your job:

- **List job run**: You can run the following command, which uses the `list-job-run` option to list the jobs with their state information:

```
aws emr-containers list-job-runs --virtual-cluster-id
<cluster-id>
```

- **Describe a job run**: You can run the following command, which uses the `describe-job-run` option to learn more about the job, which includes the job state, state details, and job name:

```
aws emr-containers describe-job-run --virtual-cluster-id
<cluster_id> --id <job-run-id>
```

- **Cancel a job run**: You can run the following command, which uses the `cancel-job-run` option to cancel the running jobs:

```
aws emr-containers cancel-job-run -virtual-cluster-id
<cluster_id> --id <job-run-id>
```

Now, let's take a look at the next deployment option.

Amazon EMR on AWS Outposts

AWS Outposts is a fully managed service that gives you access to the same AWS services, infrastructure, APIs, and operational models as virtually any data center or on-premises facility. AWS Outposts is great for workloads that require low-latency access by keeping infrastructure near to the data center.

AWS services related to compute, storage, or databases run locally on Outposts and you can access these services available in your AWS Region to build and scale your on-premises applications using the same AWS tools and services.

There is a range of AWS services, including AWS compute, storage, and databases, that run locally on AWS Outposts. Amazon EMR is also available in AWS Outposts, which allows you to deploy, manage, and scale Hadoop and Spark workloads in your on-premises environments similar to as you would do in the cloud.

Using the same AWS console, SDK, or CLI commands as EMR, you can easily create a managed EMR cluster in your on-premises environment, and these clusters running in AWS Outposts will be available in the AWS console the same as other clusters.

There are a few prerequisites that you need to follow to use Amazon EMR on
AWS Outposts:

- You need to have an AWS account.
- You must have installed and configured AWS Outposts in your on-premises
  infrastructure or data center.
- You will need a stable network connectivity between your Outposts environment
  and your selected AWS Region.
- You need to have enough capacity of EMR-supported instance types in
  your Outposts.

For connectivity, you can extend your AWS account's VPC to span its AZs to associated
Outposts locations. While creating an EMR cluster, you should configure your Outposts
to be associated with a subnet that extends your regional VPC environment to your
on-premises deployment.

The following is an example AWS CLI command to create an EMR cluster in Outposts and
it's pretty much the same as creating a cluster in the AWS cloud with EC2:

```
aws emr create-cluster \
--name "Outpost cluster" \
--release-label emr-<label> \
--applications Name=<app-names> \
--ec2-attributes KeyName=<key-name> SubnetId=subnet-<id> \
--instance-type <type> --instance-count <count> --use-default-
roles
```

Please replace the <label>, <app-names>, <key-name>, subnet <id>, instance
<type>, and instance <count> variables with the relevant values before executing it.

With Amazon EMR in AWS Outposts, you will get all the benefits of Amazon EMR with
the following additional benefits:

- **Easier integration with on-premises deployments**: Workloads running on
  Amazon EMR on AWS Outposts can read from and write data to your existing
  on-premises Hadoop cluster's HDFS storage. This gives you the flexibility to
  implement your data processing needs using Amazon EMR without migrating
  any data.

- **Accelerate the data and job migration**: If you are in the process of planning to migrate your on-premises Hadoop cluster data and workloads to Amazon EMR in the cloud, then as an interim step, you can start using EMR through AWS Outposts. This will allow integration with your on-premises Hadoop deployment, and then you can plan to gradually move your cluster data to S3 followed by jobs to be executed in the cloud. This way, you can get all the benefits of decoupling compute with storage.

Please check the AWS documentation to look for limitations and current support around different instance types you can select while creating your cluster. At the time of writing this book, EC2 spot instances are not available when you deploy EMR on Outposts.

EMR pricing for different deployment options

Similar to other AWS services, Amazon EMR's pricing also follows a pay-as-you-go model. You can easily estimate your costs based on the deployment option, the Region, the instance types you are selecting, and how long you plan to keep the cluster running. The pricing is calculated per-second with a 1-minute minimum billing period.

A cluster with 20 nodes running for 10 hours will cost the same as a 100-node cluster running for 2 hours. Now, you might consider always running a higher-number node cluster to finish the job in 1 minute, but that's not the ideal way of execution as Hadoop/ Spark workloads give maximum performance at a certain number of nodes and don't perform better beyond that point with any additional nodes.

As explained earlier, EMR provides three deployment options: EMR on Amazon EC2, EMR on Amazon EKS backed by EC2 or AWS Fargate, and EMR on AWS Outposts. Now, let's look at a few pricing examples for each deployment option, which can help you estimate the cost of your Hadoop/Spark workloads in EMR.

> **Important Note**
> The costs in USD represented in the examples are based on the pricing we had while writing this book and they are subject to change.

Amazon EMR on Amazon EC2 pricing

The Amazon EMR on Amazon EC2 pricing is very simple and takes pretty much the same calculation approach as Amazon EC2 instance pricing, where the pricing varies by the type of EC2 instance you have selected, the number of instances, and the size of the EBS volume attached to them.

The same as EC2 pricing, apart from **On-Demand** instance types, you can choose to go for **Reserved Instances** or **Savings Plans** or choose to use **Spot Instances** for your task nodes to get higher savings.

Please refer to the AWS documentation to see the instance types supported by EMR, as they might change from time to time.

Pricing example for EMR on EC2

To keep the calculation simple, let's take the following assumptions:

- You will be deploying the cluster in us-east-1.
- You will have one master node of C4.2xlarge, two core nodes of m5.2xlarge, and five task nodes of m5.4xlarge, and all the instances are on-demand instances.
- The cluster is up and running for 2 hours for a Spark job or step.

The following explains the cost breakdown and the total cost you will have:

- Formula to calculate each node type's cost: (number of instances) x (selected instance type's hourly cost) x (number of hours)
- EMR master node cost: 1 instance x $0.105 per hour x 2 hours = $0.21
- EMR core node cost: 2 instances x $0.096 per hour x 2 hours = $0.384
- EMR task node cost: 5 instances x $0.192 per hour x 2 hours = $1.92
- Total cost = $0.21 (master node cost) + $0.384 (core node cost) + $1.92 (task node cost) = $2.514

Let's now understand the pricing of another deployment option.

Amazon EMR on Amazon EKS pricing

When you are considering deploying an Amazon EMR cluster on Amazon EKS, you have the option to select Amazon EKS backed by Amazon EC2 or AWS Fargate. When the Amazon EKS cluster is built on top of EC2 instances, the pricing calculation is the same as EMR on EC2 as the cost will be dependent on the type of instance, the number of instances, and the EBS volumes attached to them.

When you choose to go with EKS on AWS Fargate, pricing is calculated based on the vCPUs and the amount of memory used from the time you start downloading the container image until the EKS Pod terminates, and the time is rounded up to the nearest second as the pricing is per-second billing. With AWS Fargate, pricing is based on the amount of vCPU cores and memory used by the Pod.

You can check the EMR pricing page to find out how much is charged per vCPU core and memory per GB.

Estimating costs for EMR on EKS with AWS Fargate

This is a simple formula you can use to calculate the cost of your workload:

- Total cost for vCPU usage = (number of vCPUs) * (per vCPU-hours rate) * (job runtime in hours)

- Total cost for memory usage = (amount of memory used) * (per GB-hours rate) * (job runtime in hours)

- Total cost = total cost for vCPU usage + total cost for memory usage

Please note, apart from the vCPU core and memory usage cost, you pay an additional $0.10 per hour for each Amazon EKS cluster you have launched. You can use the same EKS cluster for multiple workloads, where you create separation between workloads through Kubernetes namespaces and AWS IAM security policies.

The per-vCPU core hourly rate and per-GB memory rate used in the following example is taken from the EMR on EKS pricing page. You can select the AWS Region you plan to deploy and get the defined cost at the time of implementation.

> **Important Note**
> The pricing calculation formula and the cost specified here are subject to change and are based on prices at the time of writing this book. Please refer to the AWS documentation for the latest pricing information.

Pricing example for EMR on EKS with AWS Fargate

Let's assume that the EKS cluster is deployed in the N. Virginia (us-east-1) Region and you have used the r5.xlarge EC2 instance type, which has 4 vCPU cores and 32 GB memory.

If we create the EKS cluster with 100 nodes or instances, then we have a total of 400 vCPU cores and 3,200 GB memory capacity.

Now, let's assume we have a Spark application running in the cluster and that takes 100 vCPU cores and 500 GB memory and the job executed for 1 hour; then, you can apply the formula we specified previously to arrive at the cost:

- Total cost for vCPU usage = 100 * $0.01012 * 1 = $1.012

- Total cost for memory usage = 500 * $0.00111125 * 1 = $0.5556

- Total cost = $1.012 (total cost for vCPU usage) + $0.5556 (total cost for memory usage) + $0.10 (EKS per hour cost) = $1.667

Let's understand the pricing for another deployment option.

Amazon EMR on AWS Outposts pricing

Before getting into the pricing for AWS Outposts, let's see what AWS Outposts offers.

AWS Outposts helps you deploy an Amazon EMR cluster near to your on-premises environment, making it a part of your existing environment but the features, its usage, and the AWS APIs are all the same as the AWS cloud. Amazon EMR on AWS Outposts provides a cost-efficient option with the same benefits of automating different administration tasks of provisioning infrastructure resources, setting up the cluster, and configuring Hadoop libraries or tuning.

Coming back to the pricing of Amazon EMR on AWS Outposts, it's the same as in the cloud. Please refer to the AWS Outposts pricing page of the AWS documentation for more details.

Monitoring and controlling your costs with AWS Budgets and Cost Explorer

Apart from the pricing considerations, you can also take advantage of AWS Budgets and Cost Explorer to monitor your cluster costs and also set some alarms to send notifications to your finance team. Following are a few examples on how to use these applications to your advantage:

- Use AWS Budgets to configure custom budgets that will help track your cost and usage.

- When real or anticipated costs and usage exceed your budget level, and actual **Reserved Instances** and **Savings Plans** utilization or coverage falls below the desired thresholds, you can choose to be notified through email or SNS.

- Use AWS Cost Explorer to view and analyze your cost and usage drivers.

- AWS Budgets integrates with AWS Service Catalog, which enables you to track costs on AWS services that are approved by your organization.

With this, you have a good overview of the different EMR deployment options and how the pricing is calculated for each one of them. Please refer to the *Further reading* section for additional learning materials related to pricing.

Summary

Over the course of this chapter, we have dived deep into the Amazon EMR architecture, each of its components, and Hadoop applications. After covering those topics, we then discussed cluster nodes of EMR with its life cycle and ways to submit jobs.

Finally, we covered what the different EMR deployment options are, what benefits they have, and what the pricing for each of them is.

That concludes this chapter! Hopefully, you have got a good overview of the Amazon EMR architecture with its three deployment options and are ready to learn different use case architecture patterns in the next chapter.

Test your knowledge

Before moving on to the next chapter, test your knowledge with the following questions:

1. You receive a daily incremental file from your source system at midnight and you are expected to process it and make it available for consumption. After that, during the day, at 3 P.M., you need to execute a machine learning job that will read this processed output. Will you use a persistent cluster or transient and how will you configure it?

2. While creating an EMR cluster, you have a requirement to select multiple instance types for your node types and would like to take advantage of spot instances too. How would you configure your cluster?

3. You have a manufacturing unit that expects all the Hadoop/Spark processing to happen near its on-premises site, but has plans to slowly migrate to the cloud. Which Amazon EMR deployment option is best suited?

Further reading

The following are a few resources you can refer to for further reading:

- Setting of an Amazon EMR on EKS cluster: `https://docs.aws.amazon.com/emr/latest/EMR-on-EKS-DevelopmentGuide/setting-up.html`

- Amazon EMR on AWS Outposts limitations: `https://docs.aws.amazon.com/emr/latest/ManagementGuide/emr-plan-outposts.html`

- Amazon EMR pricing: `https://aws.amazon.com/emr/pricing/`

3
Common Use Cases and Architecture Patterns

This chapter provides an overview of common use cases and architecture patterns you will see with **Amazon Elastic MapReduce** (**EMR**) and how EMR integrates with different AWS services to solve specific use cases. The use cases include batch **Extract, Transform, and Load** (**ETL**), real-time streaming, clickstream analytics, interactive analytics with **machine learning** (**ML**), genomics data analysis, and log analytics.

This should give you a starting point to understand what problem statements you can solve using Amazon EMR and use it to solve your real-world big data use cases.

We will dive deep into the following topics in this chapter:

- Reference architecture for batch ETL workloads
- Reference architecture for clickstream analytics
- Reference architecture for interactive analytics and ML

- Reference architecture for real-time streaming analytics
- Reference architecture for genomics data analytics
- Reference architecture for log analytics

Reference architecture for batch ETL workloads

When data analysts receive data from different data sources, the first thing they do is transform it into a format that can be used for analysis or reporting. This data transformation process might involve several steps to bring it to the desired state and, after it is ready, you need to load it into a data warehousing system or data lake, which can be consumed by data analysts or data scientists.

To make the data available for consumption, you need to *extract* it from the source, *transform* it with different steps, and then *load* it into the target storage layer – hence the term ETL. For a few other use cases, when the raw data is in a structured format, you can then load it into a relational database or data warehouse and then transform it with SQL, where it becomes **Extract, Load, and Transform** (**ELT**).

What we understand from all this is that *transformation* is the primary piece that makes raw data ready for consumption. What needs to be done as part of the transformation process depends on the type or format of data, any data cleaning that needs to be done, any business transformations that need to be applied or not, and many more.

In earlier days, when data volumes used to be small (in MB), you had the option to use typical programming languages and a single server to process them, but with higher volumes of data (gigabytes to petabytes), you need to look at distributed processing capability with Hadoop/Spark and execute through Amazon EMR.

Batch ETL use cases are very common in the data analytics world and the job execution can be through a scheduler or triggered based on a file arrival event. In this section, we will explain a reference architecture for batch ETL workloads that receives data from different sources and after multiple transformation steps, the transformed output is ready for data analysis or **business intelligence** (**BI**) reporting.

Use case overview

For this use case, let's assume, as an organization, you receive data from the following two sources:

- **On-premises systems**, which includes two data sources. One is a relational database and the other is a filesystem.

- A **vendor filesystem** that uses **SSH File Transfer Protocol** (**SFTP**) to send files.

You have an objective of curating the data in an Amazon **Simple Storage Service** (**S3**) data lake through different storage layers, such as raw, processed, and consumption, and then making the data available for consumption, where you should be able to access the data through SQL for analysis or build reports through a self-service BI reporting tool.

Reference architecture walkthrough

Now, to provide a technical solution to the preceding use case, you can refer to the following architecture where Amazon S3 is used as a persistent storage layer, Amazon EMR is being integrated to perform ETL transformations, and Amazon QuickSight is being used for BI reporting:

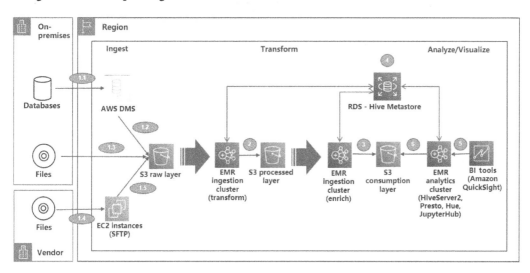

Figure 3.1 – Reference architecture for batch ETL workloads

As part of this architecture, you will notice the following three Amazon S3 storage layers:

- **S3 raw layer**: This layer is responsible for keeping the raw data from the source system as it is, which can help you to reprocess the pipeline in case of issues or if you plan to reuse the same data for other downstream processing. If you do not keep the raw data persistently, you may have to pull the data from the source system again when needed and the older data might be missing from the source system itself. So, it's always better to keep the raw data persistently separate if you need to handle future needs, such as reprocessing the data to address any earlier transformation issues or using it for another use case.

- **S3 processed layer**: After you receive raw data, you may need to apply a few common data cleansing, data validation, or standardization processes to make the data ready for consumption. This might include validating the number of records, the file size, or the schema of each record before making it available for consumption. Data engineers use this layer to apply business logic transformation and create a subset of data for different downstream systems.

- **S3 consumption layer**: This layer is the final storage layer, which is used by data analysts to analyze data to derive insights or build BI reports or dashboards for leadership teams. The processed layer data goes through transformations based on the consumption need and is written to the consumption layer with the required file format and partitioning strategies.

Here is an explanation of the architecture steps:

- *Step 1* represents data ingestion or movement from on-premises to an Amazon S3 raw bucket:

  - *Steps 1.1 and 1.2* represent pulling data from a relational database source where you can integrate AWS **Data Migration Service** (**DMS**) to pull data to an Amazon S3 bucket through a scheduled pull mechanism. AWS DMS is a managed service using which you can move data from different databases (on-premises or the cloud) to different AWS databases (Amazon **Relational Database Service** (**RDS**) or Amazon Aurora), Amazon S3, or AWS analytics services such as Amazon **Kinesis Data Streams** (**KDS**), Amazon OpenSearch Service, and Amazon Redshift. This enables you to pull data from the source as a one-time extract or pull data on a continuous basis using a **Change Data Capture** (**CDC**) mechanism.

  - *Step 1.3* represents your on-premises files that can be uploaded to Amazon S3 directly.

- *Steps 1.4 and 1.5* represent the vendor file being pushed to an SFTP server hosted on Amazon EC2, from which you can push to Amazon S3 through a scheduled batch script.

- *Step 2* integrates a transient EMR cluster that takes input from the raw S3 bucket, applies required cleansing or standardization rules, and writes the output back to the S3 processed bucket. For the ETL transformations, you can use Hive or Spark steps in EMR.

- *Step 3* applies additional business logic to enrich the processed datasets. Here also, an Amazon EMR transient cluster is integrated that reads input from the processed S3 bucket and applies business logic transformations through Spark, then writes enriched output to the consumption S3 bucket.

- *Step 4* represents an external Hive Metastore built on top of Amazon RDS, which is used by all the transient EMR clusters so that each transient cluster can refer to the existing data catalog instead of creating it from scratch.

- *Step 5* represents the consumption layer, where with EMR's Hive or Presto query engines, you can access the data from the S3 consumption bucket through SQL. You can also use the Hue web interface for data analysis.

  Then, on top of that, you can integrate Amazon QuickSight or any other BI reporting tool to create aggregated visualizations or report dashboards.

Now let's discuss a few best practices that you can follow while implementing the pipeline.

Best practices to follow during implementation

While there are use case-specific recommendations, the following are some generic guidelines you can follow while implementing this use case:

- **File formats**: When you perform ETL operations, at the final stage, you write the output back to an Amazon S3 data lake and you have different file format options to select from, for example, Parquet, Avro, ORC, JSON, and CSV.

  You might receive raw data as JSON or CSV, but for analytics use cases, columnar formats such as Parquet are very popular. Columnar formats provide storage savings and provide great performance when you query specific columns through SQL.

- **Partitioning**: This is a data distribution technique where you first identify your query patterns to understand how you filter your datasets and then use the filter columns as your partition columns. On the storage layer, it creates subfolders based on the partition column value. For example, if most of your queries are with a country column filter, then you can select a country as the partition column and in S3, you will have a country-based subfolder and each country subfolder will have records related to that country. So, when you filter by a country, you only scan that country-specific subfolder, which gives you better performance.

- **Transient EMR clusters**: As this is a batch ETL workload, your ETL jobs are mostly scheduled to run once daily or multiple times a day. When the jobs are not being executed, your cluster becomes idle and you still pay for the cluster usage time as the infrastructure is still up and running. So, for batch ETL jobs, transient EMR clusters are better suited as you save infrastructure costs when the EMR cluster resources are not being utilized.

- **External Hive Metastore**: When you have transient EMR cluster workloads, it's always recommended to go with an external Hive Metastore so that you don't lose catalog data when your cluster is getting terminated and also, you can share the catalog with multiple EMR clusters. For an external Hive Metastore, you can have an Amazon RDS database or you can use AWS Glue Data Catalog.

I hope this provides a good overview of batch ETL use cases and you will now be able to integrate Amazon EMR to build a data analytics pipeline. In the next section, you will learn how EMR can be integrated for a clickstream analytics use case.

Reference architecture for clickstream analytics

In consumer-facing applications, such as web applications or mobile applications, business owners are more interested in identifying metrics from a user's access patterns to derive insights into which products, services, or features users like more. This enables business leaders to make more accurate decisions. Often, it becomes a necessity to capture user actions or clicks in real time to have a real-time dashboard suggesting how successful your campaign is or how users are responding to your new product launch.

To make business decisions in real time, you need to have the data flow in near real time too. This means as soon as the user clicks anywhere within the application, you need to capture an event immediately and push it through your backend system for processing. As multiple users access your application through different channels, it generates a stream of events and you need a scalable architecture that can support receiving a massive volume of concurrent records and can also use a big data processing framework to process them.

Use case overview

Here, we will take an example use case to explain how you can integrate a few AWS services with Amazon EMR to build a real-time streaming clickstream application.

It assumes your organization has a website that has a lot of user traffic on a daily basis. To track overall usage patterns, you have integrated Google Analytics into your website and for detailed user session-based analytics, you plan to stream click events in real time too.

Your objective is to aggregate both Google Analytics and real-time clickstream events into a data lake and also ingest aggregated output to a data warehouse, on top of which you can build real-time BI reports.

Reference architecture walkthrough

Now, to provide a technical solution for the preceding use case, you can refer to the following architecture where Amazon S3 is integrated as a data lake, Amazon Redshift is used for a data warehouse, Amazon EMR is integrated to perform ETL and real-time streaming, and Amazon QuickSight is integrated for BI reporting:

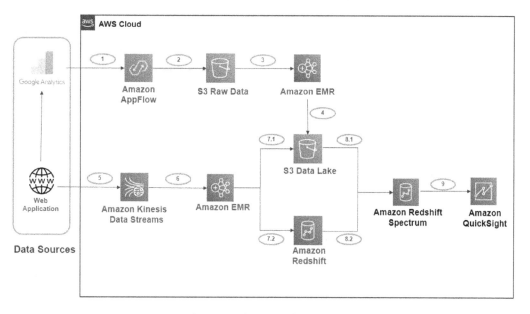

Figure 3.2 – Reference architecture for clickstream analytics

Apart from Amazon S3, EMR, and QuickSight, you will notice other AWS services integrated, which are as follows:

- **Amazon AppFlow**: Amazon AppFlow is a fully managed integration service that can help you transfer data between **software-as-a-service (SaaS)** applications, such as Salesforce, Google Analytics, and Slack, and AWS services, such as Amazon S3 and Amazon Redshift, in just a few clicks on the AppFlow interface. You can schedule these flows to run at regular intervals or integrate them to get triggered with different business events. It also provides a few configurations in terms of selecting which attributes you want to transfer, what the target storage format should be, or what the partitioning structure in Amazon S3 should be.

  For this clickstream analytics use case, you can integrate Amazon AppFlow to pull data from Google Analytics to Amazon S3 in a scheduled pull manner.

- **Amazon KDS**: Amazon KDS is a scalable and durable message bus that can help integrate real-time streaming use cases, where you can have multiple producers pushing data to KDS and there can be multiple consumers who are reading from the stream in real time. It is a serverless service that can stream gigabytes of data per second and is a great fit to stream real-time clickstreams, log events, financial transactions, or social media feed events. You can integrate multiple consumers, including AWS Lambda, EMR with Spark Streaming, or AWS Glue with Spark Streaming or Flink, to process the stream events in real time and write to the target storage.

- **Amazon Redshift**: Amazon Redshift offers massively parallel processing capability and is great for high-volume data warehousing platform needs. Its high performance makes it popular in the analytics world and is commonly integrated as a backend database for real-time BI reports.

Here is an explanation of the architecture steps:

- *Steps 1 and 2* represent data movement from the Google Analytics tool to an Amazon S3 raw bucket with the help of Amazon AppFlow. You can define a scheduled pull from Google Analytics and store the output JSON in S3. Please note, the output you receive is a nested JSON that requires additional transformation to flatten it for consumption.

- *Steps 3 and 4* integrate a transient EMR cluster that takes nested JSON input from the raw S3 bucket, flattens it, applies minimal cleansing or standardization rules, and writes the output back to the S3 data lake bucket. For the ETL transformations, you integrate a Spark step in EMR. *Step 3* applies additional business logic to enrich the processed datasets. Here also, an Amazon EMR transient cluster is integrated that reads input from processed S3 bucket and applies business logic transformations through Spark, then writes enriched output to the consumption S3 bucket.

- *Step 5* represents the web application that can integrate Amazon KDS's **Kinesis Producer Library** (**KPL**) to ingest data into KDS partitions as soon as a user click event happens. KPL also provides in-built buffering and retry mechanisms to handle failures.

- *Steps 6 and 7* represent Amazon EMR with Spark Structured Streaming as the consumer application of the Kinesis stream. It does two operations; one is writing the raw events to Amazon S3 and the second is aggregating the stream data with Google Analytics and writing aggregated output to the Amazon Redshift data warehouse.

- *Steps 8 and 9* represent the consumption layer, where we have integrated Amazon Redshift Spectrum to query data from both the data lake and Amazon Redshift storage layers and then integrate Amazon QuickSight on top of it to build real-time reports or dashboards. This is the layer where your business users join to see how your campaigns perform in real time or how your users are reacting to your new product launch.

Now let's discuss a few best practices that you can follow while implementing the pipeline.

Best practices to follow during implementation

Here are a few generic guidelines that you can follow while implementing this use case; for sure there will be more use case-specific ones:

- **Scalability**: You don't need to invest in creating a massive cluster for your KDS or Amazon EMR cluster from day 1. You can use the scaling features available in both of them to scale the cluster up and down as the volume of stream events changes throughout the day. Amazon EMR provides in-built autoscaling and managed scaling features that you can use to scale your cluster in real time.

- **Fault tolerance**: You also need to consider how you recover from failures. Regarding KDS, you can use the data retention setting, which is by default set to 7 days but can be extended up to 1 year. Then, you can take advantage of EMR with Spark Streaming's checkpointing feature using which you can checkpoint stream events to the Amazon S3 location. If your EMR cluster or Spark job gets terminated, you can restart from the failure point that is checkpointed in Amazon S3.

- **Optimizing data lake storage**: As discussed in *Figure 3.1 – Reference architecture for batch ETL workloads,* you can optimize storage in a data lake by choosing the appropriate file format and also structuring your data into subfolders or partitions, which will give better performance when you query the table with the partition column as a filter.

- **Use Redshift for aggregated output**: You have the option to write the complete raw dataset into Redshift too as it's structured data but avoid doing so as Redshift infrastructure will add a lot to the cost, compared to data lake storage. So, creating a mix of a data lake and data mart is great for use cases such as this.

With this use case, you have learned how you can implement clickstream analytics using Amazon EMR and what some of the general recommendations you can follow are. In the next section, let's understand how you can do interactive analytics with Amazon EMR as a long-running cluster.

Reference architecture for interactive analytics and ML

In the previous sections of this chapter, you might have seen the usage of Amazon EMR as a transient cluster that gets created through file arrival or a scheduled event, processes the file with Hive or Spark steps, and then gets terminated. Transient clusters are great to decouple storage and compute and also to save costs by reducing cluster idle time.

But there are few use cases where you might need a persistent EMR cluster that might be active 24x7 with minimal cluster node capacity and goes through the EMR autoscaling feature to scale up and down as needed. These persistent clusters generally serve multiple workloads, including ETL transformations with Hive/Spark, analyzing data through SQL-based query engines such as Hive and Presto, or interactive ML model development through notebooks. In a few cases, you can implement a multi-tenant EMR cluster that serves multiple teams with an access policy and data isolation.

As the cluster is available 24x7, multiple users use the same cluster compute capacity for different workloads. Then, you can configure the cluster to define queues that will have required CPU and memory resources assigned that get used through EMR's capacity and fair schedulers.

Use case overview

Here, we will see an example to explain how a persistent EMR cluster can serve multiple workloads and make the data available to data analysts and data scientists for interactively querying the datasets or ML model development.

It assumes your organization receives data from two different sources that need to be aggregated into a data lake. Then, the aggregated output should be available to your data analysts for interactive querying using SQL or your data scientists for exploring the datasets and doing ML model development, model training, and inference.

To give an overview of ML engineering, it goes through a sequence of steps. Initially, data scientists put effort into exploring the datasets, cleansing or preparing them, and identifying the attributes that make the most sense for model development. Then, they start ML model development using the Python or R scripting languages using different ML frameworks, such as TensorFlow, MXNet, or PyTorch. After the model development is ready, they train the model on historical datasets and optimize their model as needed.

The model training process creates trained models that will be used to predict output against new input datasets. So, in general, model development and training is an intensive process where data scientists go through a lot of iterations to standardize the data, optimize ML model code, and do training with a proper mix of datasets.

Now let's understand the two data sources we have, which will be the input to the data pipeline:

- An external vendor is sending a daily CSV file to your input S3 bucket directly that includes the financial credit score of your customers, which needs to go through a validation and cleansing process.

- You have subscribed to another vendor's data feed that exposes data through REST APIs and you need to pull the data dynamically from their APIs in a periodic fashion. After extracting data, you need to flatten the results and store them in your data lake.

Let's take a look at the reference architecture.

Reference architecture walkthrough

Now, to provide a technical solution for the preceding use case, you can refer to the following architecture, where we have the following:

- Amazon S3 is integrated for the data lake.

- Amazon EMR is used for both transient and persistent EMR clusters.

- AWS Lambda and Amazon DynamoDB are integrated to automate data extraction from REST APIs.

- The EMR cluster's Hive, Presto, Zeppelin, or Jupyter notebooks are integrated to provide an interactive development experience for data scientists.

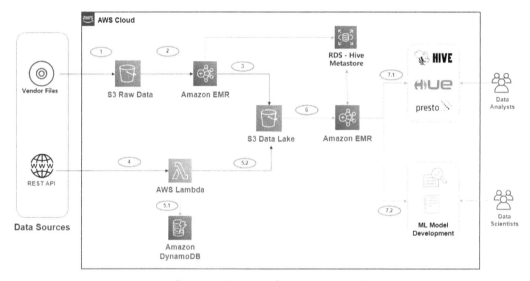

Figure 3.3 – Reference architecture for interactive analytics and ML

Before going deep into the architecture, let's understand the two new AWS services we have introduced as part of this architecture:

- **AWS Lambda**: AWS Lambda is a serverless compute service that lets you run code without provisioning or managing any infrastructure. It provides flexibility to scale compute capacity as needed and it also allows you to choose from a variety of programming languages for implementing your business logic. You can use AWS Lambda's user interface to write code or package your code as a ZIP file and upload it to AWS Lambda for execution. You can integrate AWS Lambda to get triggered through various events or through a scheduler and also, it natively integrates with other AWS services, such as Amazon API Gateway, Amazon DynamoDB, Amazon KDS, and AWS Step Functions, to build your application in a decoupled approach.

For this use case, you can integrate AWS Lambda to get triggered at a regular interval (for example, every 5 minutes or every 1 hour) to connect to the REST API, fetch datasets, transform it, and write to a target Amazon S3 data lake.

- **Amazon DynamoDB**: Amazon DynamoDB is a key-value NoSQL database in AWS that delivers single-digit millisecond performance at any scale. It is fully managed and can be deployed as a global database across multiple regions of AWS. It is often used as the backend of scalable REST APIs or as a key-value metastore.

   For this use case, DynamoDB is integrated to keep track of the Lambda execution and the amount of data being pulled from the REST API. So, if the Lambda execution fails, it can refer to the DynamoDB table to identify the failure point and trigger again from that point.

- **Notebooks**: Amazon EMR integrates both Jupyter and Zeppelin notebooks, which you can configure to submit jobs and queries to the Amazon EMR cluster's Hadoop/ Spark interfaces.

   For this use case, analysts can use a notebook to execute their queries or jobs step by step during development.

Here is an explanation of the architecture steps:

- *Steps 1-3* represent the vendor file ingestion to the data lake. *Step 1* represents the vendor directly pushing the file to your input raw S3 bucket. Then, you can integrate a transient EMR cluster on top of it, which might have a Hive or Spark job step to read from the input bucket, apply required transformations, and write the final output to the S3 data lake bucket.

- *Steps 4 and 5* represent ingesting REST API data to the data lake. You can schedule the AWS Lambda function to be executed every 30 minutes, which might be written in Python. It will connect to the REST API, get the response as JSON, apply a few transformations to flatten it, and then finally, write the transformed output to the data lake. As explained earlier, we need to bring in a mechanism to recover from failures as the REST APIs will have throttling enabled to restrict frequent access to the APIs. *Step 5.1* represents writing metadata to a DynamoDB table, where every time the Lambda function pulls data from the REST API, it will checkpoint which timestamp or record ID it pulled so that the next execution will be from the earlier checkpointed timestamp.

- *Step 6* represents a persistent Amazon EMR cluster that users connect to explore the data available in the data lake. EMR provides the distributed processing capability with all Hadoop ecosystem services such as Hive, Spark, Presto, and Jupyter Notebook for interactive analytics. It also integrates Amazon RDS as its external Hive Metastore.

- *Step 7.1* integrates Hadoop interfaces, such as Hue, Hive, and Presto, which data analysts can use to analyze the data in the data lake through SQL. Hue is a web interface that acts as a client for users, which integrates with the Hive catalog and can submit queries to query engines such as Hive and Presto.

  Hive is a distributed query engine that can be configured to submit queries to MapReduce or Spark. It will parse the user-submitted queries to MapReduce or a Spark-equivalent script, which will read input data from the data lake, do processing, and then serve the result in tabular format.

  Similar to Hive, Presto is a low-latency query engine that you can use to execute ad hoc analytical queries.

- *Step 7.2* represents the integration of notebooks on top of the EMR cluster for interactively developing Python, PySpark, R, and other scripts to analyze data, do ML model development, or train your model for inference.

  EMR provides options to select Jupyter Notebook or Zeppelin notebooks on top of the EMR cluster, which comes with different scripting language options, and you can also integrate your own plugins and modules for development.

Now, let's discuss a few best practices that you can follow while implementing the pipeline.

Best practices to follow during implementation

Here are a few generic guidelines that you can follow while implementing this use case; for sure there will be more use case-specific ones:

- **Cluster resource management**: As this is a persistent EMR cluster that will be used for multiple workloads, you need to have a strategy to manage cluster resources well. You can think of defining different queues to manage high-priority and low-priority jobs so that when high-priority jobs are getting executed, low-priority jobs can wait for their completion.

  You can configure multiple queues with different amounts of memory and CPU resources as per your needs and configure the capacity scheduler or fair scheduler to let the scheduler decide how to respond as multiple workloads get executed concurrently.

  After you have defined queues, you can specify `-queue <queue-name>` as an additional parameter for your `spark-submit` commands to direct YARN to use the specified queue for this job.

- **Cluster capacity planning and scaling**: You can monitor your cluster usage for a period of time and derive patterns around the minimum cluster capacity you need on a continuous basis and keep that as your minimum cluster capacity. On top of that, you can configure EMR's autoscaling or managed scaling features to scale up or down as new workloads come in.

  In addition, you also need to consider the amount of HDFS space you need on the cluster, depending on your implementation. Even if you have integrated Amazon S3 as your persistent data store, you may still need to cache some amount of data in HDFS for better performance.

- **Data isolation and security**: When you have a persistent cluster being shared by multiple teams and multiple workloads, you also need to make sure the security aspects are integrated so that you are able to configure authentication, authorization, and encryption of data at rest or during transit.

  For authenticating users' access to your cluster, you can create a mix of AWS IAM and your Active Directory integration.

  For authorization, you can integrate AWS Lake Formation or Apache Ranger, where you can define which user can access which catalog databases, tables, or columns.

To make your data secure at rest, you can enable encryption with AWS **Key Management Service (KMS)** keys or custom keys, and for making the data secure while it's in transit, make sure you have SSL/TLS integrated.

Reference architecture for real-time streaming analytics

At the beginning of the chapter, you learned about clickstream analytics that integrated Amazon KDS and EMR with Spark Streaming to stream clickstream events in real time. The use case covered in this section is another use case that explains how you can stream **Internet of Things (IoT)** device events in real time to your data lake and data warehouse for real-time dashboards.

To give an overview of IoT, it's a network of physical objects called *things* that uses sensors and related software technologies to connect and exchange information with other devices or systems over the internet. These devices can be any household or industrial equipment that has a sensor and required software embedded into it to communicate with other devices or send messages to a central unit that monitors requests or signals.

The adoption of IoT around the world is increasing as analytics on device data can provide a lot of insights to optimize usage or predict patterns.

Use case overview

Let's assume your organization has IoT devices to track electric usage at anybody's home or office. Your plan is to sell these IoT devices and help set them up at your customers' homes or offices. This will track all usage of electricity and help stream the data in real time to a centralized data lake and data warehouse in AWS.

Your organization's data analysts will analyze these real-time datasets, aggregate them with historical data to derive insights, and then provide analytical reports to their users.

These analytical reports might include usage patterns around which days of the week or which time of the day they consume more electricity or which electronic devices in their homes consume more electricity. Your organization also might provide recommendations around how your customers can save their monthly costs by changing their usage patterns.

As the IoT devices will stream every bit of device usage information in real time, it is expected that the data volume will be higher, and to handle processing such a bigger volume of data, you need a big data processing service or tool such as Amazon EMR.

Reference architecture walkthrough

Now, to provide a technical solution for the preceding use case, you can refer to the following architecture where AWS IoT is integrated to receive IoT events and publish them to KDS. Then, Amazon EMR helps in further aggregations to make the aggregated data available in Redshift and an Amazon S3 data lake for analytics:

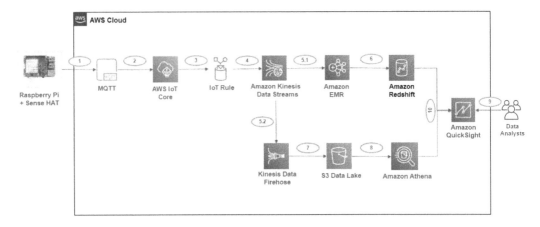

Figure 3.4 – Reference architecture for real-time streaming analytics

Before going deep into the architecture, let's understand the three new AWS services we have introduced as part of this architecture:

- **AWS IoT**: AWS IoT provides capability using which you can connect one IoT device with another and connect your IoT devices to the AWS cloud. It provides device software that will help you integrate your IoT device with AWS IoT-based solutions. It has several components, such as AWS IoT Core, FreeRTOS, AWS IoT Greengrass, AWS IoT 1-Click, AWS IoT Analytics, AWS IoT Button, AWS IoT Device Defender, AWS IoT Device Management, AWS IoT Events, AWS IoT SiteWise, AWS IoT Things Graph, and AWS Partner Device Catalog.

  AWS IoT Core provides support for **Message Queuing and Telemetry Transport (MQTT)**, MQTT over **WSS (WebSockets Secure)**, **Hypertext Transfer Protocol Secure (HTTPS)**, and **Long Range Wide Area Network (LoRaWAN)**, which gives a wide range of flexibility for integration.

  AWS IoT natively integrates with other AWS services, which can help you to implement an end-to-end pipeline faster.

- **Kinesis Data Firehose**: Amazon Kinesis Data Firehose is a fully managed service with scalability built in, which provides delivery stream capability that can deliver streaming data to Amazon S3, the Amazon OpenSearch service, Amazon Redshift, HTTP endpoints, and a few third-party service providers, such as Splunk, Datadog, and New Relic.

  It provides additional features, such as the buffering of stream messages, applying transformations through AWS Lambda, and delivering to Amazon S3 with different file formats or partitioning in place.

  Kinesis Data Firehose is popular for delivering KDS messages to its supported targets with minimal transformations in near real time. For this IoT use case, it does something similar where it reads IoT events from KDS and writes them back to Amazon S3.

- **Amazon Athena**: Amazon Athena is an interactive query engine that is built on top of Presto and uses Apache Hive for **Data Definition Language** (**DDL**) internally. It is serverless, which means there is no infrastructure to manage, and this also follows the pay-as-you-go pricing model. Athena's pricing is based on the amount of data you are scanning.

  Athena is very popular for querying data lakes or Amazon S3, where you define a virtual table on top of an Amazon S3 path, add required partitions to the table as needed, and then execute standard SQL queries to get results. It adds a lot of value when analysts are more familiar with SQL-based analysis compared to complex ETL programming.

  Athena is integrated with AWS Glue Data Catalog out of the box. Also, apart from querying from Amazon S3, Athena also supports querying from other relational or third-party data sources through its Federated Query feature, which uses AWS Lambda internally to fetch data from the source and provide results to Athena in a tabular format.

The following is an explanation of the architecture steps:

- *Steps 1 and 2* represent the IoT devices sending electricity usage metrics to AWS IoT Core through MQTT in real time.

- *Steps 3 and 4* represent AWS IoT Core using IoT rules to send event messages to KDS, which will facilitate multiple consumer applications to read data from KDS and write it to multiple targets.

  You should be considering what the different types of events you are going to receive from your IoT devices are and define partition keys in KDS accordingly.

- *Steps 5.1 and 6* represent one of the consumers of the KDS events, where Amazon EMR uses Spark Structured Streaming to read data from KDS in real time and write to Amazon Redshift. In this case, the EMR cluster will be an always-on persistent cluster with a minimal number of nodes to stream data on a continuous basis.

- *Steps 5.2, 7, and 8* represent another consumer of KDS where Kinesis Data Firehose is integrated to define a delivery stream with an Amazon S3 data lake as the target and on top of which Amazon Athena is integrated for querying the data through standard SQL. The purpose of writing the data to a data lake is to have a persistent data store for the historical data, whereas Amazon Redshift is being used as a data mart to store aggregated output for real-time reporting.

- *Steps 9 and 10* represent the integration of Amazon QuickSight on top of an S3 data lake and Redshift data mart for building BI reports. QuickSight can use Amazon Athena to query from an S3 data lake or use Redshift Spectrum to read data from both S3 and Redshift to show a real-time report on aggregated output.

Now let's discuss a few best practices that you can follow while implementing the pipeline.

Best practices to follow during implementation

Here are a few generic guidelines that you can follow while implementing this use case; for sure there will be more use case-specific ones:

- **Buffering and partitioning configuration of Kinesis Data Firehose**: Kinesis Data Firehose has configurations where you can specify whether you would like to buffer the data before delivering it to Amazon S3, which might help in aggregating a number of records to a single file in Amazon S3 to avoid too many small files in S3. Please note, too many small files (files in KBs or less than 64 MB) in an S3 data lake might create a performance bottleneck when you try to access them through Amazon Athena, Amazon Redshift, or Redshift Spectrum as it will create a lot of overhead in tracking so many small files. You can take advantage of the buffering configuration of Kinesis Data Firehose and also consider partitioning configuration while writing data back to an S3 target.

- **Distribution and sort key of Redshift**: As you write data into a Redshift cluster, it gets distributed across compute nodes so that when you submit queries, it can execute your query in a distributed fashion. But Redshift provides flexibility to choose how you would like to distribute your queries across nodes so that your join or filter queries can perform better. It provides the **Key**, **Even**, and **All** distribution styles to choose from.

  In addition to the distribution key, Redshift also provides flexibility to select a *sort* key, which helps to decide in which order data will be sorted internally. When the data is sorted, it enables the Redshift query optimizer to scan fewer chunks of data, which in turn will give higher performance. Redshift provides two types of sort keys: a compound sort key and an interleaved sort key.

Now that we have understood this use case, let's dive into another in the following section.

Reference architecture for genomics data analytics

Before going into the technical implementation details of genomics data analytics, let's understand what **genomics** means. It is a field of study of biology that focuses on the evolution, mapping, structure, and functions of genomes. A genome is a complete set of DNA of a living being, which includes all of its genes.

In recent times, there have been significant investments in genomics and clinical data to explore more about living beings' genes and their characteristics, which can help diagnose any disease beforehand or predict new features. Technology continues to play a vital role in genomics studies: as the data volume grows, you can use big data technologies for distributed processing.

Genomics datasets are available in complex data formats, such as VCF and gVCF, and to parse them, there are several popular frameworks available, such as Glow and Hail.

Use case overview

Let's assume your organization is providing products or services that can help in detecting, diagnosing, or treating different health diseases, and for this, your organization heavily invests in genomics studies. Your organization has its own research data and gets data genomics and clinical data from third-party vendors to aggregate it with in-house data and derive insights out of it.

You are in need of a big data processing solution for the genomics clinical data and also a centralized data store. For the whole solution, you plan to use AWS cloud-native services.

Reference architecture walkthrough

Now, to provide a technical solution for the preceding use case, you can refer to the following architecture, where you can use Amazon EMR with Spark and one of the open source VCF file processing frameworks, such as Glow, for the ETL need. Apart from EMR's big data processing capability, you can use Amazon S3 for the persistent storage of historical data, Amazon Redshift as a data warehouse for aggregated data, and Amazon QuickSight for BI reporting:

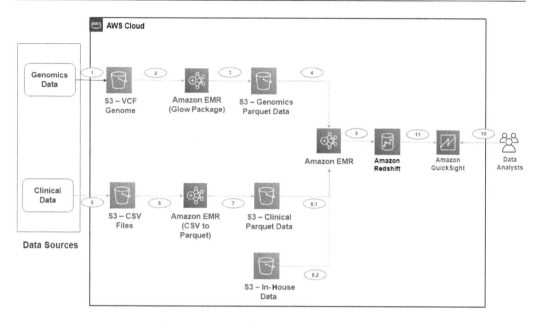

Figure 3.5 – Reference architecture for genomics data analysis

Before going deep into the architecture, let's understand a bit about the open source Glow package, which you can integrate into an EMR Spark job.

Glow is an open source utility that helps you work with genomics data and is built on Apache Spark, which can help in big data processing. It supports processing popular genomics formats, such as VCF and BGEN, and can scale with Spark's distributed processing. Because of its native Spark support, you have the flexibility to choose from the Spark SQL, Python, Scala, or R languages.

The following is an explanation of the architecture steps:

- *Steps 1, 2, and 3* represent the data ingestion pipeline for genomics data that you receive from your external vendor. *Step 1* represents your vendor writing genomics VCF files to your input S3 bucket directly and your objective is to parse them and convert them to standard Parquet format.

  Steps 2 and 3 represent a transient EMR cluster that is scheduled to run a Spark job every day, which will read input data from the input S3 bucket, parse it through the Glow package, and write the Parquet to the output bucket with a daily partition (year/month/day). It moves the input files to a processed folder after successful transformation, so the next execution picks up the new files.

- *Steps 5, 6, and 7* represent the data ingestion pipeline for the clinical data that you receive from another vendor. This is a simple pipeline compared to the genomics data pipeline, which reads input CSV files, converts them to Parquet, and writes output to an S3 target bucket with day-wise partitions. Here also, you can integrate a transient EMR cluster that runs a Spark job every day at midnight.

- *Steps 4, 8.1, 8.2, and 9* represent another transient EMR cluster, whose responsibility is to aggregate all data sources, derive aggregate output, and push it to Amazon Redshift for further analytics or reporting needs. Apart from the vendor's genomics and clinical data, we also have in-house datasets in another S3 bucket. A scheduled EMR job reads data from all three input buckets and does transformations and aggregations using Spark.

- *Steps 10 and 11* represent the integration of Amazon QuickSight on top of an Amazon Redshift data warehouse for building BI reports.

Now let's discuss a few best practices that you can follow while implementing the pipeline.

Best practices to follow during implementation

Here are a few generic guidelines that you can follow while implementing this use case; for sure there will be more use case-specific ones:

- **Bootstrap action in EMR for additional library configuration**: As discussed, there are several frameworks or utilities that support parsing genomics file formats. We have taken the example of Glow here, but you can integrate others, such as Hail. When you are in need of configuring external libraries in your EMR cluster, use its bootstrap actions, which will configure the clusters as needed while launching the cluster. When you have transient EMR cluster use cases like this, these bootstrap actions will help a lot as every time a cluster is launched, it will automatically configure external libraries in the required nodes.

- **Distribution and sort key of Redshift**: As discussed earlier, choose from the **Key**, **Even**, and **All** distribution styles for your data in Redshift depending on your query pattern or join scenarios. Also, see whether you need to use sort keys to let the Redshift optimizer choose fewer chunks of data while querying.

Now, let's look at the reference architecture for log analytics in detail.

Reference architecture for log analytics

Log analytics is a common requirement in most enterprises. As you grow with multiple applications, jobs, or servers that produce enormous logs every day, it becomes essential to aggregate them for analysis.

There are several challenges in log analytics as you need to define log collection mechanisms, process them to apply common cleansing and standardizations, and make them available for consumption. Each server or application produces its own format for logs and your job is to bring them to a format that you can use and use technologies to handle the heavy volume of log streams.

Use case overview

Let's assume your organization is on AWS and you have multiple applications deployed on AWS EC2 instances. These applications are written in Java and a few other languages and hosted through Apache or NGINX servers. You have the following three log streams that are generating logs continuously, which you plan to collect and make available for consumption:

- **EC2 server logs**: Each of your EC2 servers is generating logs that include CPU, memory usage, error logs, or access logs.

- **Application logs**: Each application is generating debug or error logs. For example, Java applications are generating logs through the Log4j framework.

- **Apache or NGINX server logs**: When applications are deployed or accessed through Apache or NGINX servers, they also generate access logs or error logs.

There are different teams in your organization that are interested in accessing these logs and they have their own preferred tools to access them. The following are the consumers:

- **Security team**: Your security team is collecting logs from different sources and is interested in EC2 access logs to make sure there is no unauthorized access and also that a hacking attack is not happening. They use Splunk as their tool for log analytics and would like to get these EC2 access logs into Splunk too.

- **DevOps team**: Your DevOps team is more interested in getting all software configuration information and also CPU and memory usage of the EC2 instances in real time to react to additional resource provisioning. Your DevOps team has their own Redshift cluster, on top of which they do further analysis and reporting. So, they expect the logs to be pushed to their Redshift cluster.

- **Application team**: The application team is more interested in analyzing their application logs to find common failure patterns. They prefer loading the application logs into an **Amazon OpenSearch Service** cluster, where they would like to do a regular expression or pattern matching search on the last 3 months of data.

Apart from these three teams, your organization has a requirement to archive all logs in Amazon S3 to compliance requirements.

Reference architecture walkthrough

Now, to provide a technical solution for the preceding use case, you can refer to the following architecture where all servers and applications will be publishing logs to KDS and EMR with a Spark Streaming job that can parse the logs and send them to the respective target depending on the log type.

Please note, all four targets we have here (Splunk, Redshift, Amazon ES, and S3) are also supported on Kinesis Data Firehose. The question may arise why we are not integrating Kinesis Data Firehose to read from KDS and write to the defined target—and, yes, that can be done—but EMR provides the following benefits over Kinesis Data Firehose:

- EMR with Spark Streaming will provide real-time streaming, compared to the near real-time streaming of Kinesis Data Firehose.

- You have different types of logs coming in, which might come through different partition keys of KDS. With EMR, you will get the flexibility to loop data by topic, apply respective transformation rules, and write to the target. If you use Kinesis Data Firehose, each delivery stream will receive all the data and you will have to integrate additional Lambda functions to filter by topic.

- As the number of log types and different targets increases, you will have to define more delivery streams by type. So, it may not scale in the future.

The following diagram shows the reference architecture of log analytics:

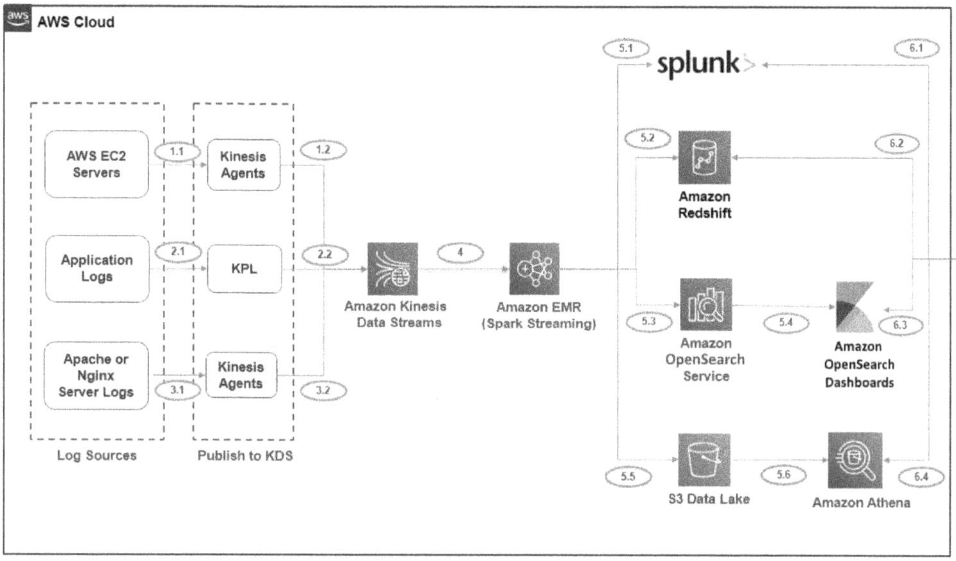

Figure 3.6 – Reference architecture for log analytics

Before going deep into the architecture, let's understand a bit about Amazon OpenSearch and Splunk, which are a couple of new components introduced in this architecture:

- **Amazon ES**: This is a fully managed service that facilitates the easy setup and deployment of the open source OpenSearch service at scale. You can do everything that you do with the open source OpenSearch service with native integration with other AWS services and cloud security built in. In our use case, it facilitates faster search with pattern matching. So, when you receive millions of log records every day, searching through them and finding patterns becomes key and the Amazon OpenSearch service is great at it.

- **Splunk**: This is a software platform, commonly used for log analytics or search use cases, that captures machine-generated data and indexes it for faster search. It also supports generating alerts, graphs, visualization dashboards, and so on. For this use case, we have assumed your organization's security team analyzes access logs using Splunk.

The following is an explanation of the architecture steps:

- *Steps 1.1 and 1.2* represent publishing EC2 server logs to KDS, where you can set up and configure a Kinesis agent in each of the EC2 servers that will read logs from the server log file path and push to KDS with any optional buffering configurations.

- *Steps 2.1 and 2.2* represent applications using the **Kinesis Producer Library** (**KPL**) to publish logs dynamically to KDS. For a few logging frameworks, you have native integrations to submit logs instantaneously to KDS using KPL.

- *Steps 3.1 and 3.2* represent publishing application server logs using the same Kinesis agent approach that is being integrated for EC2 server logs.

- *Step 4* represents the integration of Amazon EMR with Spark Streaming, which will read from KDS and write to different targets, depending on the log type or schema. Please note, as you have different log types with different schemas being ingested to the same KDS, you need a way to separate them. The best method you can follow is defining partition keys by log type in KDS and letting EMR loop through each partition key and define the target by log type.

  This way, your EMR Spark Streaming code can decide whether the log type is an access log, then write to Splunk, and if it is an Apache log, then write to Amazon OpenSearch.

- *Steps 5.1 and 5.2* represent an EMR Spark Streaming job writing to Splunk and Amazon Redshift.

- *Steps 5.3 and 5.4* represent an EMR Spark Streaming job writing to Amazon ES for faster search, and then the integration of an Amazon OpenSearch Dashboard visualization on top of it for reporting dashboards.

- *Steps 5.5 and 5.6* represent writing all the log types to Amazon S3 and accessing them through Amazon Athena's standard SQL query. Please note, when you write to S3, define folders or buckets based on log type and also use partitioning for better performance.

- *Steps 6.1-6.4* represent the consumption layer, where different teams will use their respective tool's interface to access the data. Teams using Splunk and Redshift can access their console or APIs to access the data, whereas teams using Amazon OpenSearch can use Amazon OpenSearch Dashboard to access data or build visualizations on top of it.

Now let's discuss a few best practices that you can follow while implementing the pipeline.

Best practices to follow during implementation

Here are a few high-level guidelines that you can follow while implementing this use case:

- **Configure source log types to partition the key and target**: The major challenge we have here is publishing all types of logs with different schemas to one KDS. The first thing you need to define is what the different types of logs are that should be treated separately and map them to different partition keys of KDS so that your Spark Streaming consumer application can read by partition key and write to different targets.

  The other thing you can do is use **AWS Glue Schema Registry**, which you can integrate with KDS to enforce a schema on your data. A schema of a record represents the format and structure of the data and AWS Glue Schema Registry helps to enforce the schema and provide a centralized place to manage, control, and evolve your schema.

- **Scaling KDS and EMR clusters**: You can use EMR's autoscaling or managed scaling features to scale your cluster up and down as the log stream data volume changes.

  To scale KDS, you can take advantage of CloudWatch metrics, which will provide the read and write metrics of each shard of the Kinesis stream, and based on that, you can have your custom application or AWS Lambda function that will add or remove shards from your KDS cluster.

With this last use case, you have got a good overview of different use cases you can implement with Amazon EMR that includes transient and persistent EMR clusters.

Summary

Over the course of this chapter, we have dived deep into a few common use cases where Amazon EMR can be integrated for big data processing. We discussed how you can integrate Amazon EMR as a persistent or transient cluster and how you can use it for batch ETL, real-time streaming, interactive analytics, and ML and log analytics use cases. Each use case explained a reference architecture and a few recommendations around its implementation.

That concludes this chapter! Hopefully, you have got a good overview of different architecture patterns around Amazon EMR and are ready to dive deep into different Hadoop interfaces and EMR Studio in the next chapter.

Test your knowledge

Before moving on to the next chapter, test your knowledge with the following questions:

1. Assume you are receiving data from multiple data sources and after ETL transformation storing the historical data in a data lake built on top of Amazon S3 and storing aggregated data in the Redshift data warehouse. You have a requirement to provide unified query engine access, where your users can join both data lake and data warehouse data for analytics. How will you design the architecture and which query engine you will recommend to your analysts?

2. Your organization has multiple teams and departments that have different big data and ML workloads. They plan to use a common EMR cluster that they can use for their analytics and ML model development. Your data scientists are new to Amazon EMR and would like to understand how they can take advantage of this EMR cluster to do ML model development. What will your guidance be?

3. You have a customer use case where you need to stream IoT device events into a data lake and data warehouse for further analysis. Because of cost constraints, your customer team is ready to compromise the real-time streaming requirement and is happy to wait for 5 minutes to 1 hour for the data to arrive. How will you design the architecture so that it's cost-efficient and at the same time solves the business problem?

Further reading

The following are a few resources you can refer to for further reading:

*   Different EMR case studies (search for EMR): `https://aws.amazon.com/solutions/case-studies/`

*   Redshift distribution style: `https://docs.aws.amazon.com/redshift/latest/dg/c_best-practices-best-dist-key.html`

*   Read more about AWS DMS: `https://aws.amazon.com/dms/`

*   Read more about AWS IoT: `https://aws.amazon.com/iot/`

4
Big Data Applications and Notebooks Available in Amazon EMR

From previous chapters, you got an overview of **Amazon EMR (Elastic MapReduce)**, its architecture, and reference architecture for a few common use cases. This chapter will help you learn more about a few of the popular big data applications and distributed processing components of the **Hadoop** ecosystem that are available in EMR, such as **Hive**, **Presto**, **Spark**, **HBase**, **Hue**, **Ganglia**, and so on. Apart from that, it will also provide an overview of a few machine learning frameworks available in EMR, such as **TensorFlow** and **MXNet**.

At the end of the chapter, you will learn about notebook options available in EMR for interactive development that include EMR Notebook, **JupyterHub**, **EMR Studio**, and **Zeppelin** notebooks.

The following topics will be covered in this chapter:

- Understanding popular big data applications in EMR
- Understanding machine learning frameworks available in EMR
- Understanding notebook options available in EMR

Technical requirements

In this chapter, we will cover different big data applications available in EMR and how you can access or configure them. Please make sure you have access to the following resources before continuing:

- An AWS account
- An IAM user, which has permission to create EMR clusters, EC2 instances, and dependent IAM roles

Now let's dive deep into each of the big data applications and machine learning frameworks available in EMR.

Understanding popular big data applications in EMR

There are several big data applications available in the Hadoop ecosystem and open source community, and EMR includes a few very popular ones that are very commonly used in big data use cases. The availability of different big data applications or components in your cluster depends on the EMR release you choose while launching the cluster. Each EMR release includes a different version of these applications and makes sure they are compatible with each other for smooth execution of the cluster and jobs.

EMR does include the most common or popular Hadoop interfaces in its recent releases and also does continuous updates to include new Hadoop interfaces as they gain popularity in the open source community. In addition to adding new big data applications or components, EMR also removes support for a few as they lack attention from the open source community or customers. For example, till EMR 3.11.x, you had the option to select **Impala** as an application, but after 4.x.x, support for Impala was removed.

To explain the different big data components available in EMR, we considered the latest EMR release available while writing this book, which is 6.3.0.

Now let's look at a few of the popular components available in EMR, such as Hive, Presto, Spark, HBase, Hue, and Ganglia, which are pretty commonly used in big data use cases. You will also learn about how these open source components are configured in EMR and how they are integrated with Amazon S3.

Hive

Hive is an open source query engine that runs on top of Hadoop and allows you to query data from your data lake or cloud object store with a standard SQL-like language called **Hive Query Language (Hive QL)**.

Compared to typical relational databases, Hive follows schema on read semantics instead of schema on write. That means you can define a table schema on top of your HDFS or Amazon S3 file path, which is called as a virtual table, and when you run a Hive query on top of your metadata table, it will fetch data from your underlying storage, apply the schema on top of it, and show the output in tabular format.

Hive supports reading from different types of file formats, such as CSV, JSON, Avro, Parquet, ORC, and so on, with the inclusion of the respective file format's serializer. You can also bring in your custom serializer and specify that in your Hive table properties to let Hive know how to parse your file format.

Hive internally uses big data processing frameworks such as MapReduce, Tez, and Spark to fetch data from HDFS or S3 and process it to serve the output. It converts user-submitted Hive QL to corresponding MapReduce, Tez, or Spark jobs and you can configure Hive to use any of these frameworks.

Passing variables to Hive Step in EMR

You can use AWS's EMR Console or AWS CLI to trigger a Hive Step. It also supports passing variables to a Hive script that you can access using a $ sign and curly braces, for example, ${variable-name}.

For example, if you would like to pass an S3 path to your Hive script with a variable name of MyPath, then you can pass it as the following:

```
-d MyPath=s3://elasticmapreduce/lookup-input/path1
```

Then you can access it in your Hive script as ${MyPath}.

Additional considerations for Hive with Amazon S3 as a persistent storage layer

When Hive is configured on top of the EMR core node's HDFS, its way of working is the same as non-EMR environments, but there are slight variations when it is configured on top of Amazon S3 as its persistent storage layer.

The following are a few of the differences to consider:

- **Authorization**: EMR supports Hive authorization only for HDFS and not for S3 and EMRFS. Hive Authorization is disabled in an EMR cluster by default.

- **File merge behavior**: Apache Hive provides two configurations to control if you need to merge small files at the end of a map-only job. The first configuration is `hive.merge.mapfiles`, which needs to be true, and the second configuration is to trigger the merge if the average output size is less than the value set in the `hive.merge.smallfiles.avgsize` parameter. Hive on EMR works with both these settings if the output is set to HDFS but if the output is set to Amazon S3, then the merge task is always triggered if `hive.merge.mapfiles` is set to `true`, ignoring the value set for `hive.merge.smallfiles.avgsize`.

- **ACID transactions**: Hive **ACID** (**Atomicity, Consistency, Isolation, Durability**) transaction support is available in EMR from EMR 6.1.0 and above.

- **Hive Live Long and Process** (**LLAP**): Hive 2.0 included a new feature called Live Long and Process, which provides a hybrid execution model. It uses long-lived daemons that replace interaction with HDFS and also provides in-memory caching that improves performance. This Hive feature is only available after the EMR 6.0.0 release.

These differences or considerations are explained assuming you will be using EMR 6.x.x. If you plan to go with older EMR releases, there might be other differences that you should consider.

Integrating an external metastore for Hive

As explained, Hive defines a virtual table or schema on top of file storage to facilitate SQL-like query support on your data. To store all the table metadata, Hive needs to use a persistent metastore. By default, Hive uses a MySQL-based relational database to store metadata, which is deployed on the master node. But this brings a risk of losing the metastore if your master node's file system gets corrupted or you lose the instance itself.

To secure the Hive metastore, you can think of externalizing the metastore, which means instead of storing it in master node's MySQL instance, look for options to store it outside the cluster, so that you can persist it to support transient EMR cluster use cases too.

In AWS, you have the following options for externalizing your Hive metastore.

- An Amazon Aurora or Amazon RDS database
- The AWS Glue Data Catalog (supported by EMR 5.8.0 and later versions only)

Now let's understand how you can configure a Hive metastore with these two options.

Configuring Amazon Aurora or RDS as a Hive metastore

Hive has `hive-site.xml`, which has configurations to specify which metastore to use. To use Amazon RDS or Amazon Aurora, you will need to create a database there and override the default configuration of `hive-site.xml` to point to this newly created database.

The following steps can guide you with the setup:

1. Create an Amazon Aurora or Amazon RDS database by following the steps given in the AWS documentation (`https://aws.amazon.com/rds/`).

2. You need to allow access between your database and the EMR master node. To do that, please modify your database cluster security group.

3. After your database is available with the required access for connectivity, you need to modify the `hive-site.xml` configuration file to specify the JDBC connection parameters of your database. To avoid modifying the original `hive-site.xml` file, you can create a copy of the file and have a new name such as `hive-config.json` with the following JSON configuration:

```
[{
        "Classification": "hive-site",
        "Properties": {
            "javax.jdo.option.ConnectionURL":
    "jdbc:mysql://<hostname>:3306/
    hive?createDatabaseIfNotExist=true",
            "javax.jdo.option.ConnectionDriverName": "org.
    mariadb.jdbc.Driver",
            "javax.jdo.option.ConnectionUserName":
    "<username>",
```

```
        "javax.jdo.option.ConnectionPassword":
    "<password>"
        }
    }]
```

In this configuration file, you need to replace `<hostname>`, which will be your database server's host, and `<username>` and `<password>`, which will be your database connection credentials. We have specified `3306` as the port number, assuming you have a MySQL database with the default port, but you can change it as needed.

4. After your configuration file is ready, the next step is to specify the file path while creating your EMR cluster. The configuration file path can be a local path or an S3 path.

   The following is an example of an AWS CLI command to launch an EMR cluster:

   ```
   aws emr create-cluster --release-label emr-6.3.0
   --instance-type m5.xlarge --instance-count 2
   --applications Name=Hive --configurations file://hive-
   config.json --use-default-roles
   ```

As you will notice, we have `hive-config.json` specified as the configuration file for Hive. We have referred to a local path here, but you can upload this configuration JSON to S3 and use the S3 path.

Configuring the AWS Glue Data Catalog as a Hive metastore

As explained earlier, AWS Glue is a fully managed ETL service, which is built on top of Spark and has Glue Crawler, Glue Data Catalog, Glue Jobs, and Glue Workflows as primary components. The AWS Glue Data Catalog provides a unified metadata repository, which can be shared across multiple AWS services, such as Amazon EMR, AWS Lake Formation, Amazon Athena, Amazon Redshift, and so on.

Starting with EMR 5.8.0 and later releases, you can configure the AWS Glue Data Catalog as a Hive external metastore, which can be shared across multiple EMR clusters, Glue Spark jobs, or even can be shared with multiple AWS accounts.

When you integrate Glue Data Catalog as EMR's external metastore, you need to consider the Glue Data Catalog pricing too. The Glue Data Catalog provides storage for up to 1 million objects for free every month and beyond that, for every 100,000 objects, you will be charged USD 1 each month. An object in Glue Catalog is represented as a database, table, or partition.

Now let's understand how to configure Glue Data Catalog as Hive's external metastore in EMR:

- **Through the AWS Console**: If you are creating the EMR cluster through the AWS console, then create the cluster through **Advanced Options**, select EMR 5.8.0 or a later release, choose Hive or the HCatalog service under the release, and then under **AWS Glue Data Catalog settings**, choose **Use for Hive table metadata**. You can select the rest of the options as needed and proceed with the cluster creation. This should set you up to use AWS Glue Data Catalog as an external Hive metastore.

- **Through the AWS CLI**: If you are going to create the cluster through AWS CLI commands, then you can specify the `hive.metastore.client.factory.class` value using the `hive-site` classification. The following is an example of the configuration:

```
[{
    "Classification": "hive-site",
    "Properties": {
        "hive.metastore.client.factory.
class": "com.amazonaws.glue.catalog.metastore.
AWSGlueDataCatalogHiveClientFactory"
    }
}]
```

- **Additional configuration for specific EMR releases**: If you are using EMR release 5.28.0, 5.28.1, or 5.29.0, then you need to specify additional configuration in the `hive-site` configuration, where you need to set `hive.metastore.schema.verification` as `false`. If this is not set to `false`, the master instance group will be suspended.

- **Configuring Glue Catalog available in other AWS accounts**: If your Glue Data Catalog is in another AWS account, then you will have to specify an additional configuration where you need to specify the account ID in the `hive.metastore.glue.catalogid` parameter. The following is an example of the JSON configuration:

```
[{
    "Classification": "hive-site",
    "Properties": {
        "hive.metastore.client.factory.
class": "com.amazonaws.glue.catalog.metastore.
AWSGlueDataCatalogHiveClientFactory",
```

```
        "hive.metastore.schema.verification": "false"
        "hive.metastore.glue.catalogid": "<account-id>"
    }
}]
```

A few additional IAM permission configurations might be needed if you are not using the default `EMR_EC2_DefaultRole` role with the `AmazonElasticMapReduceforEC2Role` managed policy attached to it, and also if you have additional encryption or decryption procedures involved. Please refer to the AWS documentation link specified in the *Further reading* section of this chapter.

Presto

Similar to Hive, Presto also provides a distributed query engine to query data from different data sources such as HDFS, Amazon S3, and Kafka and databases such as MySQL, MongoDB, Cassandra, Teradata, and so on with a SQL-like query language. But compared to Hive's batch engine, Presto provides a high-performance fast SQL query engine designed specifically for interactive query use cases.

Presto is available in two separate versions, PrestoDB and PrestoSQL. Presto was originally created by a few members at Facebook and later it got forked to a separate open source release with the name PrestoSQL, which was recently renamed as Trino. In EMR, the Presto name refers to PrestoDB.

While launching your EMR cluster, you need to select either PrestoDB or PrestoSQL as selecting both is not supported. Please refer to the AWS documentation to understand the PrestoDB or PrestoSQL version attached to each release of EMR.

Both PrestoDB and PrestoSQL can access the data in Amazon S3 through the **EMR File System** (**EMRFS**). PrestoDB can access EMRFS starting with EMR release 5.12.0 and is also specified as the default configuration. PrestoSQL also uses EMRFS as the default since EMR release 6.1.0.

Making Presto work with the AWS Glue Data Catalog

As was discussed about Hive in the previous section, you can also configure Presto to use the AWS Glue Data Catalog as its external metastore. The following steps will guide you to configure it:

- **Through the AWS Console**: If you are creating the EMR cluster through the AWS console, then create the cluster through **Advanced Options**, select EMR 5.10.0 or a later release, and choose **Presto** as the application under the release. Then choose the **Use for Presto table metadata** option and select **Next** to proceed with the rest of the configuration as needed for your cluster creation.

- **Through the AWS CLI**: PrestoSQL and PrestoDB have different configurations for different EMR releases. Coming up are a few of the configuration examples.

  For PrestoDB in EMR release 5.16.0 or later, you can use the following JSON configuration to specify `"glue"` as the default metastore:

```
[{
    "Classification": "presto-connector-hive",
    "Properties": {
      "hive.metastore": "glue"
    }
}]
```

  Similar to Hive, you can also specify Glue Catalog in another AWS account with `hive.metastore.glue.catalogid`. The following is an example of the JSON configuration:

```
[{
    "Classification": "presto-connector-hive",
    "Properties": {
      "hive.metastore": "glue",
      "hive.metastore.glue.catalogid": "acct-id"
    }
}]
```

PrestoSQL started supporting Glue as its default metastore starting with the EMR 6.1.0 release. The following JSON example shows how you can specify `"glue"` in the `"prestosql-connector-hive"` configuration classification:

```
[{
    "Classification": "prestosql-connector-hive",
    "Properties": {
      "hive.metastore": "glue"
    }
}]
```

In this section, you have learned how you can integrate Presto in EMR with a few of the configurations that make it work with the AWS Glue Data Catalog. In the next section, you will get an overview of Apache Spark and its integration with Amazon EMR.

Spark

Apache Spark is a very popular distributed processing framework that supports a wide range of big data analytics use cases, such as Batch ETL with Spark Core and Spark SQL, real-time streaming with Spark structured streaming, machine learning with MLlib, and graph processing with its GraphX library. Its programming interfaces are available in Java, Scala, Python, and R, which drives its adoption.

Spark provides an in-memory distributed processing capability on top of the data stored in HDFS, Amazon S3, databases connected through JDBC, other cloud object stores, and additional caching solutions such as Alluxio. It has a **Directed Acyclic Graph** (**DAG**) execution engine that is optimized for fast performance.

You can set up or configure Spark on an EMR cluster as you do for other applications or services. Spark on EMR natively integrates with EMRFS to read from or write data to Amazon S3. As highlighted earlier, you can configure Hive to submit queries to Spark for in-memory processing.

The latest EMR 6.3.0 release includes Spark 3.1.1 and you can refer to the AWS documentation to find which specific version of Spark is included in which EMR release.

In the following example AWS CLI command, you will learn how you can create an EMR cluster with Spark as the selected service. The command is the same as you have seen for Hive or Presto:

```
aws emr create-cluster --name "EMR Spark cluster" --release-
label emr-6.3.0 --applications Name=Spark --ec2-attributes
KeyName=<myEC2KeyPair> --instance-type m5.xlarge --instance-
count 3 --use-default-roles
```

You will have to replace <myEC2KeyPair> with your EC2 key pair name before executing this command in the AWS CLI.

Making Spark SQL work with the AWS Glue Data Catalog

As discussed for Hive and Presto in previous sections, you can also configure Spark to use the AWS Glue Data Catalog as its external metastore. The following steps will guide you on how you can configure it:

- **Through the AWS Console**: If you are creating the EMR cluster through the AWS console, then create the cluster through **Advanced Options**, select EMR 5.8.0 or a later release, then choose Spark or Zeppelin as the applications under the release. Then, under **AWS Glue Data Catalog settings**, select the **Use for Spark table metadata** option and proceed with the rest of the configuration as needed for your cluster creation.

- **Through the AWS CLI**: If you are creating an EMR cluster with the AWS CLI or SDK, then you can specify the Glue Data Catalog option in the hive.metastore. client.factory.class parameter of the spark-hive-site classification.

  The following is an example of the JSON configuration:

```
[{
    "Classification": "spark-hive-site",
    "Properties": {
      "hive.metastore.client.factory.
class": "com.amazonaws.glue.catalog.metastore.
AWSGlueDataCatalogHiveClientFactory"
    }
}]
```

As explained for Hive and Presto, if you need to specify Glue Data Catalog as available in another AWS account, then in the configuration JSON, you can specify it through the "hive.metastore.glue.catalogid": "account-id" additional parameter.

Submitting a Spark job to an EMR cluster

Similar to other Hadoop services, you can submit a Spark step while launching an EMR cluster or after the cluster is created, and you can use the AWS console, the AWS CLI, or SDKs to submit a step.

Now let's see how to submit a Spark job in EMR.

Submitting a Spark job through the AWS console

Follow these steps to submit a Spark job through the AWS console to an existing EMR cluster:

1. Navigate to the Amazon EMR service console within the AWS console.
2. From the cluster list, choose the EMR cluster against which you plan to submit a job.
3. Navigate to the **Steps** section and select the **Add Step** action button.
4. In the **Add Step** dialog box, select **Spark Application** for **Step type**. Then, give a name to the step and select **Deploy mode** as **Client** or **Cluster**. Selecting client mode will launch the driver in the cluster master node, whereas cluster mode will select any node of the cluster.

    Then specify the `spark-submit` options, application script location, arguments to the script, and **Action on failure**, where you can go with the default option to **Continue**. Then choose the **Add** button, which will show the Spark job step in the steps list with the status as **Pending**.
5. Then, as the job starts running, it will move the status to **Running** and then **Completed**.

After understanding how you can submit a Spark job step to the EMR cluster using the EMR console, next let's learn how you can do the same using the AWS CLI.

Submitting a Spark job through the AWS CLI

You can submit a Spark job while launching the cluster or to an existing EMR cluster.

The following is an example of an AWS CLI command that explains how you can add a SparkPi step while launching an EMR cluster:

```
aws emr create-cluster --name "EMR Spark Cluster" --release-
label emr-6.3.0 --applications Name=Spark \
--ec2-attributes KeyName=myKeyPairName --instance-type
m5.xlarge --instance-count 3 \
```

```
--steps Type=Spark,Name="Spark Program",
ActionOnFailure=CONTINUE,Args=[--class,org.apache.spark.
examples.SparkPi,/usr/lib/spark/examples/jars/spark-examples.
jar,10] --use-default-roles
```

Alternatively you can also add a Spark step to an existing EMR cluster, as shown in the following example:

```
aws emr add-steps --cluster-id <cluster-id> --steps Type=Spark,
Name="Spark Pi Step", ActionOnFailure=CONTINUE,Args=[--
class,org.apache.spark.examples.SparkPi,/usr/lib/spark/
examples/jars/spark-examples.jar,10]
```

Please replace `<cluster-id>` with your existing EMR cluster's ID.

Improving EMR Spark performance with Amazon S3

Amazon EMR offers features and configurations using which you can improve Spark performance while reading from or writing data to Amazon S3. S3 Select and the EMRFS S3-optimized committer are a couple of methods using which you can improve the performance.

Let's understand both in a bit more detail:

- **S3 Select**: This is one of the features of S3, where you can fetch a subset of the data from S3 by applying filters on the data. When you are using Spark with EMR and trying to fetch data from S3, instead of transferring the complete S3 file to EMR and then applying a filter through Spark, you can push the filtering part to S3 Select so that less data is transferred to EMR for processing.

  It's useful when you filter out more than 50% of your data from S3, and please note that you will need a sufficient transfer speed and available bandwidth over the internet to transfer data between EMR and S3, as the data that gets transferred is uncompressed and the size might be larger. The following is sample PySpark code if you need to integrate S3 Select with Spark:

  ```
  spark
    .read
    .format("s3selectCSV") // "s3selectJson" for Json
    .schema(...) // optional, but recommended
    .options(...) // optional
    .load("s3://path/to/my/datafiles")
  ```

There are several limitations when you integrate S3 Select. A few of the limitations are that the S3 Select feature is only supported with CSV and JSON files and uncompressed or gzip files and is not supported in multiline CSV files. Please refer to the AWS documentation for a detailed list.

- **EMRFS S3-optimizer committer**: When you are using Spark with Spark SQL DataFrames or Datasets to write output to Amazon S3, the EMRFS S3-optimizer committer improves performance. The committer is available in EMR starting with the 5.19.0 release and is available by default since the 5.20.0 release.

  To enable the committer in the EMR 5.19.0 release, you need to set the `spark.sql.parquet.fs.optimized.committer.optimization-enabled` property value to `true` and you can do that by adding it to SparkConf or passing it as an argument to your `spark-submit` command.

  The following example shows how to pass it through a Spark SQL command:

  ```
  spark-sql --conf spark.sql.parquet.fs.optimized.
  committer.optimization-enabled=true -e "INSERT OVERWRITE
  TABLE new_table SELECT * FROM old_table;"
  ```

Please note this committer takes a small amount of memory for each file written by a task but that is negligible. But if you are writing a large volume of files, then the total additional memory consumed might be noticeable and in that case, you may need to tune Spark executor memory parameters to provide additional memory. In general, the guidance is if a task is to write around 100,000 files, then it might need an additional 100 MB of memory for the committer.

HBase

HBase is a popular Hadoop project of the Apache Software Foundation, which acts as a non-relational or NoSQL database in the Hadoop ecosystem. It is a columnar database, where you need to define column families and within each column family, a set of columns. In terms of architecture, HBase has master and region servers where each region server has multiple regions. In EMR, region servers will be primarily on core nodes as HDFS is configured only on core nodes.

HBase has Zookeeper built into it to provide centralized high-performance coordination between nodes or region servers. Zookeeper is an open source coordination service for distributed applications, where you can focus on your application logic and Zookeeper takes care of coordinating with hosts of the cluster by keeping metadata of all the configuration parameters.

Every time data is added, modified, or deleted, HBase keeps track of the changes as change files and then merges them periodically. This process is called compaction. HBase supports two types of compaction. One is major compaction, which you need to trigger manually as HBase does not invoke it automatically and you can define your schedule to trigger it. The other is minor compaction, which HBase does periodically without your manual intervention.

HBase also integrates with Hive where you need to define an external table on top of HBase with Hive to HBase column mapping and query the data using Hive QL.

EMR 6.3.0 has the HBase 2.2.6 version included in it. You can refer to the AWS documentation to find the HBase version included in each EMR release.

From the following example of an AWS CLI command, you can understand how you can create an EMR cluster with HBase as the selected service. The command is the same as you have seen for other Hadoop applications:

```
aws emr create-cluster --name "EMR HBase cluster" --release-
label emr-6.3.0 --applications Name=HBase --use-default-
roles --ec2-attributes KeyName=<myEC2KeyPair> --instance-type
m5.xlarge --instance-count 3
```

You will have to replace <myEC2KeyPair> in the preceding command with your EC2 key pair.

In EMR, HBase can run on top of local HDFS or Amazon S3 and it can use Amazon S3 as its root directory or use it to store HBase snapshots. HBase integration with Amazon S3 opens up several other use cases such as cross-cluster data sharing, bringing in more reliability to data storage, disaster recovery, and so on.

Let's dive deep into a few of these additional integration benefits that you can get from HBase integration with Amazon S3.

Using Amazon S3 as the HBase storage mode

Starting with the EMR 5.2.0 release, you can integrate HBase on top of Amazon S3 where you can configure HBase to store its root directory, HBase stores files, and table metadata directly in S3. With Amazon S3 as the persistent data store, you can size your EMR cluster only for compute needs, instead of considering 3x storage with the default HDFS replication factor.

This opens up support for transient EMR cluster use cases, where a cluster can be terminated after its job is completed and again start the cluster by just pointing HBase to its existing Amazon S3 root directory. You just need to make sure that at any time, only one cluster is writing to the S3 root directory path, to avoid conflict or data corruption. But you can have a read replica cluster pointing to the same path for only read operations.

Starting from the EMR 6.2.0 release, HBase uses its `hbase:storefile` system table to track the HFile paths that are used for read operations and the table is enabled by default. Thus, you don't need to perform any additional manual configuration or data migration.

The following diagram explains HBase integration with Amazon S3:

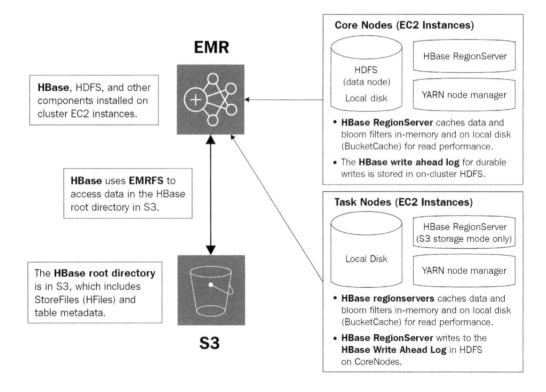

Figure 4.1 – Architecture reference for HBase on Amazon S3

Next, let's understand how you can configure HBase on your cluster to work with Amazon S3 as its persistent store.

Configuring HBase on Amazon S3 using the AWS console and the AWS CLI

You can specify configuration for HBase to work with Amazon S3 while launching the EMR cluster through the AWS console, the AWS CLI, or the AWS SDK. The following explains how you can configure it using the EMR console or the AWS CLI.

While creating the cluster through the AWS console, you can follow these steps to specify configuration for HBase on S3:

1. In the EMR console, click on **Advanced options** to create the cluster.
2. The first section of **Advanced options** you will notice is **Software Configurations**, where you can choose **HBase** or any other applications you plan to deploy.
3. Then, under **HBase Storage Settings**, select **HDFS** or **S3** and you can then select the rest of the steps as per your requirement.

If you are using an AWS CLI command to create a cluster, then please specify the following JSON configuration, where hbase.emr.storageMode will have a value of s3 and then hbase.rootdir of the hbase-site classification will point to your S3 path:

```
[
    {
        "Classification": "hbase-site",
        "Properties": {
            "hbase.rootdir": "s3://<Bucket-Name>/<HbaseStore-
Path>"
        }
    },
    {
        "Classification": "hbase",
        "Properties": {
            "hbase.emr.storageMode": "s3"
        }
    }
]
```

This section explained how you can configure HBase to work with Amazon S3. Next, you will learn what some of the HBase parameters are you can tune to get better performance.

Performance tuning parameters for HBase on Amazon S3

The following are some of the HBase configuration parameters, that you can tune to get better performance when you are using HBase on Amazon S3:

- `hbase.bucketcache.size`: This parameter represents the amount of EC2 instance store and EBS volume disk space reserved in MB for `BucketCache` storage, which is applicable to all the region server EC2 instances. By default, the value for this parameter is `8192` and a larger size value might improve performance.

- `hbase.hregion.memstore.flush.size`: This is the parameter that decides at what size or data limit (in bytes) `MemStore` will flush the cache data to Amazon S3. The default value for this is `134217728`.

- `hbase.hregion.memstore.block.multiplier`: This parameter helps HBase decide if it should block updates and look to do compaction or a `MemStore` flush. This parameter value gets multiplied by the `hbase.hregion.memstore.flush.size` value to define the upper limit, beyond which it should block updates. The default value for this parameter is `4`.

- `hbase.hstore.blockingStoreFiles`: This parameter also provides an upper limit for the maximum number of StoreFiles that can exist before blocking new updates. The default value for this is `10`.

- `hbase.hregion.max.filesize`: This represents the maximum size (in bytes) of a region before HBase decides to split the region. The default value for this is `10737418240`.

Apart from this, you can also refer to the Apache HBase documentation for other parameters that can be tuned.

Gracefully shutting down a cluster to avoid data loss

When you are using HBase with Amazon S3, it's important to shut down the cluster gracefully so that HBase flushes all `MemStore` cache files to new store files in Amazon S3. You can do that by executing the following shell script available in EMR:

```
bash /usr/lib/hbase/bin/disable_all_tables.sh
```

Alternatively, you can add a step to EMR too by using the following command:

```
Name="Disable HBase tables",Jar="command-runner.jar",Args=["/
bin/bash","/usr/lib/hbase/bin/disable_all_tables.sh"]
```

This disables all the tables, which forces each region server to flush `MemStore` cache data to S3.

Using an HBase read replica cluster

Starting from the EMR 5.7.0 release, HBase started supporting read replica clusters on Amazon S3. A single writer cluster can write to an S3 root directory and at the same time, multiple EMR read replica clusters can have read-only workloads running on top of it.

The read replica cluster is set up the same way as the primary cluster, with only one difference in the JSON configuration, which specifies the `hbase.emr.readreplica.enabled` property to be `true`.

The following is an example of the JSON configuration:

```
[
    {
        "Classification": "hbase-site",
        "Properties": {
            "hbase.rootdir": "s3://<Bucket-Name>/<HbaseStore-
    Path>"
        }
    },
    {
        "Classification": "hbase",
        "Properties": {
            "hbase.emr.storageMode": "s3",
           "hbase.emr.readreplica.enabled":"true"
        }
    }
]
```

Here, you learned how you can create an EMR HBase read replica cluster pointing to an existing S3 HBase root directory. Next, we will understand how the data gets synced while the primary cluster does write operations.

Synchronizing the read replica cluster while data is being written from the primary cluster

When you write something to HBase, it is first written to an in-memory store called `memstore` and once `memstore` reaches a certain size defined in HBase configurations, it flushes data to the persistent storage layer, which can be HDFS or an Amazon S3 layer.

When you have a read replica cluster reading from the primary cluster's HBase root directory S3 path, it will not see the latest data till the primary cluster flushes the data to S3. So to provide the read replica access to the latest data, you need to flush the data from the primary cluster more frequently, and you can do that manually or by reducing the size specified in the flush settings.

In addition to that, you will need to run the following commands in the read replica cluster to make it see the latest data:

- Run the `refresh_meta` command when the primary cluster does compaction or region split happens, or any new tables are added or removed.

- Run the `refresh_hfile` command when new records are added or existing records are modified through the primary cluster.

Using Amazon S3 to store HBase snapshots

Apart from pointing the HBase root directory to Amazon S3, you do have the option to use S3 to store a backup of your HBase table data using the HBase built-in snapshot functionality. Starting with the EMR 4.0 release, you can create HBase snapshots and store them in Amazon S3, then use the same snapshot to restore cluster data.

You can execute `hbase snapshot` CLI commands in the cluster master node and then export it to Amazon S3. You can see how you can do it using the master node's command prompt or as an EMR step in the following example.

Exporting and restoring HBase snapshots using the master node's command prompt

You can refer to the following steps to export an HBase snapshot from one cluster and restore it in another cluster using the HBase command line:

1. Create a snapshot using the following command:

    ```
    hbase snapshot create -n <snapshot-name> -t <table-name>
    ```

2. Then export the snapshot to an Amazon S3 path:

    ```
    hbase snapshot export -snapshot <snapshot-name> -copy-to
    s3://<bucket-name>/<folder> -mappers 2
    ```

3. If you have the snapshot ready, then import it into your new cluster using the following command:

```
sudo -u hbase hbase snapshot export -D hbase.
rootdir=s3://<bucket-name>/<folder> -snapshot <snapshot-
name> -copy-to hdfs://<master-public-dns-name>:8020/user/
hbase -mappers 2
```

4. After the snapshot is available in your HDFS path, you can execute the following commands, which involve disabling the table first, restoring the snapshot, and enabling it again. This is needed to avoid data corruption:

```
echo 'disable <table-name>; \
restore_snapshot snapshotName; \
enable <table-name>' | hbase shell
```

5. Please note the preceding command uses echo on the bash shell and it might still fail even if EMR returns a 0 exit code for it. If you plan to run the shell command as an EMR step, ensure you check the step logs.

In all the preceding commands, please replace the <snapshot-name>, <table-name>, <bucket-name>, <folder>, and <master-public-dns-name> variables with your input.

Exporting and restoring HBase snapshots using EMR steps

You can refer to the following steps to export and restore an HBase snapshot using EMR steps:

1. Create a snapshot using the following command:

```
aws emr add-steps --cluster-id <cluster-id> --steps
Name="HBase Shell Step", Jar="command-runner.jar", Args=[
"hbase", "snapshot", "create","-n","<snapshot-name>","-
t","<table-name>"]
```

2. Then export the snapshot to an Amazon S3 path:

```
aws emr add-steps --cluster-id <cluster-id> --steps
Name="HBase Shell Step", Jar="command-runner.jar", Args=[
"hbase", "snapshot", "export","-snapshot","<snapshot-
name>","-copy-to","s3://<bucket-name>/<folder>","-
mappers","2","-bandwidth","50"]
```

3. Then import the snapshot to your new cluster using the following command:

```
aws emr add-steps --cluster-id <cluster-id> --steps
Name="HBase Shell Step", Jar="command-runner.
jar", Args=["sudo","-u","hbase","hbase snapshot
export","-snapshot","<snapshot-name>", "-D","hbase.
rootdir=s3://<bucket-name>/<folder>", "-copy-
to","hdfs://<master-public-dns-name>:8020/user/hbase","-
mappers","2","-chmod","700"]
```

4. After the snapshot is available in your HDFS path, you need to restore the snapshot against the table for which you took the snapshot. To restore it using the AWS CLI, you can create a JSON file, which will have the following configuration, which includes the same `disable` and `enable` table commands:

```
[{
    "Name": "restore",
    "Args": ["bash", "-c", "echo $'disable
\"<tableName>\"; restore_snapshot \"<snapshot-name>\";
enable \"<table-name>\"' | hbase shell"],
    "Jar": "command-runner.jar",
    "ActionOnFailure": "CONTINUE",
    "Type": "CUSTOM_JAR"
}]
```

5. Assuming you have saved this JSON file with the name as `restore-snapshot.json`, you can add the following step to EMR to trigger the restore snapshot action:

```
aws emr add-steps --cluster-id <cluster-id> --steps
file://./restore-snapshot.json
```

In all the preceding commands, please replace the `<cluster-id>`, `<snapshot-name>`, `<table-name>`, `<bucket-name>`, `<folder>`, and `<master-public-dns-name>` variables with your input.

Hue

Hadoop User Experience (Hue) is an open source project of the Hadoop ecosystem that provides a web interface to interact with different Hadoop applications such as HDFS, Hive, Pig, Oozie, Solr, and so on. You can use your desktop system's browser to access the Hue web interface, where you can navigate through HDFS, submit queries to Hive, write Pig scripts, connect to remote databases and run queries against them, or monitor Oozie-based workflows or coordinators.

You can use Hue to act as your frontend application where you can do user management, define who can access which application, and avoid giving SSH access to your users. Your users might be data analysts or data scientists who might be interested in querying data through Hive, and they can write Hive queries, save queries, look at results in tabular format, or download a query result as CSV. They can also upload and download files through the HDFS interface or monitor Oozie workflows to track failure and restart jobs.

In EMR, Hue is installed by default when you use the **Quick Create** option in the AWS console. You can choose not to install Hue by going to the advanced options in the EMR console, or not to specify Hue as an application while using the AWS CLI. Apart from browsing HDFS, Hue in EMR does provide access to browse objects in S3 too.

> **Important Note**
> Hue in EMR does not support Hue Dashboard and PostgreSQL connectivity. Also, to access Hue Notebook for Spark, you must set up Hue with Spark and Livy.

EMR 6.3.0 includes Hue 4.9.0 and you can refer to the EMR release history in the AWS documentation to find which version of EMR includes which version of Hue.

Using Amazon RDS as a Hue database

Hue internally uses a local MySQL database hosted in EMR's master node to store its user information and query history. But you have the option to externalize the database by integrating Amazon RDS so that you can avoid data loss and can also support transient EMR cluster use cases.

To use Amazon RDS as a Hue database, you can create a configuration file in Amazon S3 pointing to the Amazon RDS database you created and use that while creating your EMR cluster.

Follow the steps given in the following section to learn how to integrate an RDS database with Hue.

Creating an Amazon RDS database using the AWS console

You can follow these steps to first create an Amazon RDS database that will be used as a Hue database:

1. Navigate to Amazon RDS in the AWS console.
2. Click **Databases** from the left navigation and then select **Create database**.
3. Then choose **MySQL** as the database engine type.

4. Then you have the option to select a Multi-AZ deployment if it's a production-critical database, which you can leave as the default and choose **Provisioned IOPS Storage**, then click **Next**. For non-production environments, you can select a single AZ as your deployment mode.

5. Then you can leave **Instance Specifications** at their defaults, specify **Settings**, and then click **Next**.

6. On the **Configure Advanced Settings** page, you need to specify the database name and a security group that allows inbound access to port 3306 from your EMR master node. If you have not created the cluster yet, then you can allow 3306 port access from all sources and restrict it after the EMR cluster is created.

7. Then click **Launch DB Instance**.

8. Next, you need to capture or save the database hostname, username, and password to connect from Hue. You can navigate to RDS Dashboard, select the instance you have created, and then, if it is available, capture all these connection credentials.

After your RDS database is ready, we can see how you can use its connection credentials with Hue, while launching your EMR cluster.

Specifying Amazon RDS for Hue while creating an EMR cluster using the AWS CLI

To specify the Amazon RDS database for Hue, the first step is to create a configuration file in Amazon S3, which will have connection credentials. Please note, I would recommend enabling S3 server-side encryption for this configuration file to keep it secure.

The following is an example of the JSON configuration file, where you can specify connection details for hue-ini classification:

```
[{
    "Classification": "hue-ini",
    "Properties": {},
    "Configurations": [
        {
            "Classification": "desktop",
            "Properties": {},
```

```
        "Configurations": [
          {
            "Classification": "database",
            "Properties": {
              "name": "<database-name>",
              "user": "<db-username>",
              "password": "<db-password>",
              "host": "<rds-db-hostname>",
              "port": "3306",
              "engine": "mysql"
            },
            "Configurations": []
          }
        ]
      }
    ]
}]
```

Please replace the `<database-name>`, `<db-username>`, `<db-password>`, and `<rds-db-hostname>` variables with your connection credentials.

Let's assume you have saved this configuration file with the name as `hue-db-config.json`. Next, you can use the following AWS CLI command to create your EMR cluster that specifies the configuration file's S3 path:

```
aws emr create-cluster --name "EMR Hue External DB" --release-
label emr-6.3.0 --applications Name=Hue Name=Spark Name=Hive
--instance-type m5.xlarge --instance-count 3 --configurations
https://s3.amazonaws.com/<bucket-name>/<folder-name>/hue-db-
config.json --use-default-roles
```

Please replace the `<bucket-name>` and `<folder-name>` variables as per your S3 path.

In this section, you have learned about Hue, how you can configure it in EMR, and how you can externalize its metastore by integrating Amazon RDS. Next, we will learn about Ganglia, which helps in monitoring your cluster resources.

Ganglia

Ganglia is an open source project that is scalable and designed to monitor the usage and performance of distributed clusters or grids. You can set up and integrate Ganglia on your cluster to monitor the performance of individual nodes and the whole cluster.

In an EMR cluster, Ganglia is configured to capture and visualize Hadoop and Spark metrics. It provides a web interface where you can see your cluster performance with different graphs and charts representing CPU and memory utilization, network traffic, and the load of the cluster.

Ganglia provides Hadoop and Spark metrics for each EC2 instance. Each metric of Ganglia is prefixed by category, for example, distributed file systems have `dfs.*` as the prefix, **Java Virtual Machine (JVM)** metrics are prefixed as `jvm.*`, and MapReduce metrics are prefixed as `mapred.*`.

For Spark, it provides metrics related to its DAGScheduler and jobs. For jobs, you can find both driver and executor metrics with a YARN application ID. As an example, they are prefixed as `application_xxxxxxxxxx_xxxx.driver.*`, `application_xxxxxxxxxx_xxxx.executor.*` and `DAGScheduler.*`. Please note, YARN based metrics are available from EMR 4.5.0 and above.

In EMR 6.3.0, Ganglia 3.7.2 version is included. You can refer to the AWS documentation to find the Ganglia version included in each EMR release.

The following is an example of the AWS CLI command that shows how you can create an EMR cluster with Ganglia as the selected service. The command is the same as you have seen for other Hadoop applications:

```
aws emr create-cluster --name "EMR cluster with Ganglia"
--release-label emr-6.3.0 \
--applications Name=Spark Name=Ganglia \
--ec2-attributes KeyName=<myEC2KeyPair> --instance-type
m5.xlarge \
--instance-count 3 --use-default-roles
```

You will have to replace `<myEC2KeyPair>` in the preceding command with your EC2 key pair.

In this section, you learned about different big data applications and how they are configured in EMR to work with other AWS services such as Amazon S3, IAM, Glue Catalog, and more. In the next section, we will provide an overview of a few of the machine learning frameworks that are available in EMR, such as TensorFlow and MXNet.

Machine learning frameworks available in EMR

There are several machine learning libraries or frameworks that you can configure in your EMR cluster. TensorFlow and MXNet are a couple of popular ones, which are available as applications that you can choose while creating the cluster.

Even though TensorFlow and MXNet are available as pre-configured machine learning frameworks in EMR, you do have the option to configure other alternatives such as PyTorch and Keras as custom libraries.

Now let's get an overview of the TensorFlow and MXNet applications in EMR.

TensorFlow

TensorFlow is an open source platform using which you can develop machine learning models. It provides tools, libraries, and a community of resources that will help researchers and data scientists to easily develop and deploy machine learning models.

TensorFlow has been available in EMR since the 5.17.0 release and the recent 6.3.0 release includes TensorFlow v2.4.1.

If you plan to configure TensorFlow in your EMR cluster, then please note that EMR uses different builds of the TensorFlow library based on the EC2 instance types you select for your cluster. For example, M5 and C5 instance types have TensorFlow 1.9.0 built with Intel MKL optimization and the P2 instance type has Tensorflow 1.9.0 built with CUDA 9.2 and cuDNN 7.1.

Using TensorBoard

TensorBoard provides a suite of visualization tools that you can use for machine learning model data exploration or experimentation. Using TensorBoard, you can track and visualize different metrics such as loss or accuracy, draw histograms, or profile your TensorFlow programs.

If you plan to configure TensorBoard in your EMR cluster, then please note that you need to start it in the EMR cluster's master node. You can refer to the following command to start TensorBoard in the master node and specify the log directory path by replacing the `<my/log/dir>` variable:

```
python3 -m tensorboard.main --logdir=</my/log/dir>
```

By default, TensorBoard uses port 6006 on the master node and you can access its web interface using the master node's public DNS. The following is the output you get in the command line after you start the service, which includes the web URL you can use.

```
TensorBoard 1.9.0 at http://<master-public-dns-name>:6006
(Press CTRL+C to quit)
```

MXNet

Apache MXNet is another popular machine learning framework that is built to ease the development of neural network and deep learning applications. Its flexible programming model with multiple languages, such as Python, Java, Scala, and R, and scalability allows for fast model training deployment.

It helps you to design neural network architectures by automating common workflows so that you can save effort on low-level computational implementations such as linear algebra operations.

Recently, MXNet started becoming more popular with its adoption across different industry use cases such as manufacturing, transportation, healthcare, and many more, with use cases related to computer vision, NLP and time series, and so on.

EMR started supporting MXNet starting in its 5.10.0 release and its recent 6.3.0 release includes the MXNet 1.7.0 version.

Notebook options available in EMR

In today's world, usage of web-based notebooks for interactive development is very common and EMR provides a few options for integrating Jupyter and Zeppelin notebooks.

Jupyter Notebook is a very popular open source web application that allows developers and analysts to do interactive development by writing live code, executing it line by line for debugging, building visualizations on top of data, and also providing narratives on code. You can also share notebooks with others, who can import code into their notebook.

Within an EMR cluster, you have the option to use EMR Notebooks and JupyterHub, and outside of your EMR cluster, you have EMR Studio, which you can attach to your EMR cluster.

Now let's dive deep into each of these options.

EMR Notebooks

EMR Notebooks is available in the EMR console. Notebooks are serverless and can be attached to any EMR cluster running Hadoop, Spark, and Livy. Using EMR Notebooks, you can open Jupyter Notebook or JupyterLab interfaces and any queries or code that you execute are instead run as a client submitting queries to your EMR on an EC2 cluster.

Your EMR Notebooks contents are saved to Amazon S3 for durability and reuse, which provides you with the option to launch a cluster, attach a notebook to the cluster for interactive development, and then terminate the cluster. As the notebook acts as a client, multiple users can have their own notebook using which they can submit queries or commands to the same EMR cluster kernel. With this feature, you don't need to configure your notebook for different EMR clusters and you can use them on-demand to save costs.

> **Important Note**
>
> Support for EMR Notebooks started from the EMR 5.18.0 release but it's recommended to use it with clusters having the latest release of 5.30.0, 5.32.0, and later or version 6.2.0 and later. There was a change made with these specific EMR releases that makes the Jupyter kernels run on the attached EMR cluster instead of the Jupyter instance, which improves performance.

There are a few limitations to consider while using EMR Notebooks:

- For EMR v5.32.0 and later, or v6.2.0 and later releases, your cluster must have the Jupyter Enterprise Gateway application running.

- EMR Notebooks works with clusters that have the `VisibleToAllUsers` setting set to `true` while creating the cluster and currently supports Spark-only clusters.

- EMR Notebooks does not support clusters that have Kerberos authentication enabled or clusters that have multiple master nodes.

- Installing custom libraries or kernels is not supported if your EMR cluster has Lake Formation permissions enabled.

Please check the AWS documentation for detailed configuration considerations and limitations.

Setting up and working with EMR Notebooks

Let's look at the following steps to guide you on how you can create an EMR notebook using the AWS console:

1. Navigate to the EMR service in the AWS console.

2. Choose **Notebooks** from the left-side navigation and then select **Create notebook**.

3. Specify your notebook name and description.

4. If you have already created an EMR cluster, then leave the default **Choose an existing cluster** option selected and click **Choose**. Then select your cluster from the list and click **Choose cluster** or select **Create cluster** and populate all fields as needed for your cluster.

5. For the final step, choose **Create Notebook**.

After your cluster is created, it goes through statuses such as **Pending**, **Starting**, and **Ready**. Once it is in the **Ready** state, you can choose **Open in Jupyter** or **Open in JupyterLab**, which will open the interface in a new tab of your browser.

Now you can select your preferred programming language, Kernel, from the **Kernel** menu and start writing, executing your code in an interactive way.

EMR Notebooks also provides a feature to execute them programmatically through EMR APIs, which allows you to pass runtime parameters that can be used as input variables in your notebook code. If you plan to execute the same code with different input variables, then this feature is very helpful as you can avoid duplicating the notebook.

The following shows a sample AWS CLI command using which you can trigger a notebook execution with a few runtime parameters:

```
aws emr --region us-east-1 \
start-notebook-execution \
--editor-id <editor-id> \
--notebook-params '{"parameter-1":"value-1", "parameter-
2":["value-1", "value-2"]}' \
--relative-path <notebook-name>.ipynb \
--notebook-execution-name <execution-name> \
--execution-engine '{"Id" : "<id>"}' \
--service-role EMR_Notebooks_DefaultRole
```

As you can see, you can pass parameters using the —notebook-params option. Please do replace <editor-id>, <notebook-name>, <execution-name>, <id>, and parameter values before executing it.

JupyterHub

JupyterHub also provides the Jupyter Notebook interface with an additional feature to host multiple instances of a single user notebook server, which creates a Docker container on the cluster master node that includes all JupyterHub components with Sparkmagic within the container.

JupyterHub includes Python 3 and Sparkmagic kernels include PySpark 3, PySpark, and Spark kernels. Sparkmagic kernels allow Jupyter Notebook to interact with the Spark service installed in your cluster using Apache Livy, which acts as a REST server for Spark. If you need to install additional kernels or libraries, you can install them manually within the container.

If you would like to list the installed libraries using conda, then you can run the following commands on your cluster master node's command line:

```
sudo docker exec jupyterhub bash -c "conda list"
```

Alternatively, you can use the following pip command too to list the installed libraries:

```
sudo docker exec jupyterhub bash -c "pip freeze"
```

Let's look at the following diagram, which explains the core components of JupyterHub in EMR with an authentication mechanism for administrators and notebook users.

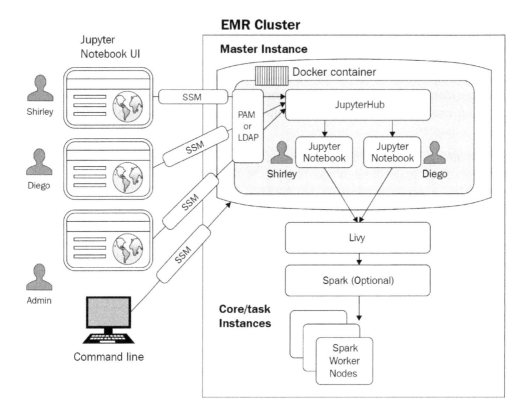

Figure 4.2 – JupyterHub architecture in EMR

EMR v6.3.0 includes JupyterHub v1.2.0 and you can refer to the AWS documentation to understand which EMR release includes which version of JupyterHub.

Setting up and configuring JupyterHub

JupyterHub is available as an application that you can choose while creating a cluster through the AWS console, the AWS CLI, or the EMR API.

While setting up JupyterHub, we need to make sure that the cluster is not created with the option to auto terminate, and the administrators and notebook users can access the EC2 key pair attached to the cluster.

The following is an example AWS CLI command, which you can use to create an EMR cluster with JupyterHub:

```
aws emr create-cluster --name="JupyterHub EMR Cluster"
--release-label emr-6.3.0 --applications Name=JupyterHub --log-
uri s3://<log-bucket>/<jupyter-cluster-logs> --use-default-
roles --instance-type m5.xlarge --instance-count 2
--ec2-attributes KeyName=<EC2KeyPairName>
```

Please replace the `<log-bucket>`, `<jupyter-cluster-logs>`, `<EC2KeyPairName>` variables before executing the command.

> **Important Note**
>
> User-created notebooks and related files are saved on the cluster's master node, which creates a risk of data loss if the cluster gets terminated. It is recommended that you have a scheduler script that continuously backs up this data.
>
> If you have done additional custom configuration changes on the container, then they will get lost if the container gets restarted. So you should have automation scripts ready that you can run to apply the custom configuration changes after the container is restarted every time.

You can also provide the following additional JSON configuration while creating your EMR cluster, which uses `jupyter-s3-conf` classification to configure JupyterHub to persist notebooks in Amazon S3:

```
[
    {
        "Classification": "jupyter-s3-conf",
        "Properties": {
            "s3.persistence.enabled": "true",
            "s3.persistence.bucket": "<jupyter-backup-bucket>"
        }
    }
]
```

With this configuration, notebooks saved by each EMR user will be saved into the `s3://<jupyter-backup-bucket>/jupyter/<jupyterhub-user-name>` path, where `<jupyter-backup-bucket>` represents the S3 backup bucket and `<jupyterhub-user-name>` represents the username of the logged-in user.

EMR Studio

EMR Studio also provides a fully managed Jupyter Notebook like EMR Notebooks but comes up with a few additional features:

- AWS Single Sign-On integration, which allows directly logging in with your corporate credentials.

- It does not need EMR console access and you can submit jobs to EMR on the EKS cluster.

- Integrates with GitHub or Bitbucket for code repository sharing.

- Provides capabilities for simpler application debugging or automating job submission to production EMR clusters using orchestration tools such as AWS Step Functions, Apache Airflow, and Amazon Managed Workflows for Apache Airflow.

Often, an organization's data engineers and data scientists do not have access to the AWS or EMR console and they would like to have their own notebook that has multiple kernels to do interactive development. For such use cases, EMR Studio is a great fit.

You can point your EMR Studio to existing EMR clusters or new ones and can also submit jobs to EMR on EKS clusters. EMR Studio adds value when it comes to building data engineering or data science applications, where you can simplify development, debugging, and deployment to production pipelines.

While setting up EMR Studio, you need to associate it with a few AWS resources such as an Amazon VPC and subnets of that VPC with a current limitation of five subnets of that VPC. As EMR Studio is associated with one VPC, you are allowed to access EMR clusters or EMR on EKS virtual clusters within that VPC and defined subnets.

EMR Studio controls its access with IAM user and permission management. Each EMR Studio instance uses a defined IAM service role and security group to provide access to an EMR cluster. It uses the IAM user role with IAM session policies to control access of an EMR Studio user.

EMR Studio is available in EMR v5.32.0 and 6.2.0 and later releases. You don't pay anything for creating or using EMR Studio and the cost is calculated based on the amount of resources you use on your EMR cluster or Amazon S3.

Workspaces in EMR Studio

A workspace is the primary building block or component of EMR Studio. The first thing you do in EMR Studio is to create a workspace, which has a similar user interface as JupyterLab and it provides additional features such as creating and attaching a workspace to EMR clusters, exploring sample notebooks, linking with GitHub or Bitbucket repositories, and executing jobs.

After a workspace is created, you can assign one of the subnets of EMR Studio to the workspace and then attach it to EMR on EC2 or EMR on an EKS cluster.

EMR Studio is associated with an Amazon S3 location and your workspace periodically autosaves the notebook cell and content to the associated Amazon S3 location between sessions. Apart from autosave, you can also manually save your notebook content with the *Ctrl + S* keys or the **Save** option under the **File** menu. Alternatively, you can also link your workspace with its repository to save it remotely and share it with your peers.

When you delete a single notebook from your workspace, its respective backup version automatically gets deleted from S3. But if you delete the workspace completely without deleting individual notebook files, then the notebook backup files does not get deleted from S3, which might add to storage costs.

Installing kernels and libraries

EMR Studio comes with pre-defined libraries and kernels but it also provides the option to install custom libraries.

The following are a couple of ways using which you can customize your EMR Studio environment when it's attached to EMR on an EC2 cluster:

- **Install Python libraries and Notebook kernels in an EMR cluster's master node**: With this option, the installed libraries will be available to all workspaces attached to the cluster. You can install libraries or kernels from the notebook itself or can SSH to your cluster's master node and install them through it.

- **Notebook-scoped libraries**: With this option, you can install libraries through your notebook cell and these notebook-scoped libraries are available to that notebook only. This option is good when some libraries are workload-specific and need not be shared across the cluster.

Please note, EMR Studio attached to EMR on an EKS virtual cluster currently does not support installing additional custom libraries or kernels.

Zeppelin

Apart from Jupyter Notebook, Apache Zeppelin also provides a web-based interactive development environment that is integrated with several interpreters, including Spark, Python, SQL, JDBC, Shell, and so on. Similar to Jupyter notebooks, you can also use Zeppelin notebooks for data ingestion, exploration, analysis, visualization, and collaboration.

Zeppelin notebooks are integrated into EMR starting with the v5.0.0 release and a few previous releases included it as a sandbox application. EMR v6.3.0 includes Zeppelin v0.9.0 and you can refer to the AWS documentation to find which version of Zeppelin is included in other releases of EMR. Starting with EMR 5.8.0, Zeppelin supports integrating AWS Glue Data Catalog as the metastore for Spark SQL. This integration is useful when you plan to persist your metadata outside of an EMR cluster or you plan to share the metastore with other EMR clusters. Please note, Zeppelin in EMR does not support SparkR Interpreter.

Summary

Over the course of this chapter, we have dived deep into a few popular big data applications available in EMR, how they are set up in EMR, and what additional configuration options or features you get when you integrate with Amazon S3. Then we provided an overview of the TensorFlow and MXNet applications, which are the machine learning and deep learning libraries available in EMR. These applications are the primary building blocks when you implement a data analytics pipeline using EMR.

Finally, we covered the different notebook options you have and how you can configure and use them for your interactive development.

That concludes this chapter! Hopefully, you have got a good overview of these distributed applications and are ready to dive deep into EMR cluster creation and configuration in the next chapter.

Test your knowledge

Before moving on to the next chapter, test your knowledge with the following questions:

1. You have terabyte-scale data available in Amazon S3 and your data analysts are looking for a query engine using which they can interactively query the data using SQL. You already have a persistent EMR cluster, which is being used for multiple ETL workloads, and to save costs you are looking for an application within EMR that can provide the interactive query engine needed. Which big data application in EMR best fits your need?

2. Your team is using EMR with Spark for multiple ETL workloads and it uses Amazon S3 as the persistent data store. For one of the use cases, you receive data that does not have a fixed schema and you are looking for a NoSQL solution that can provide data update capabilities and also can provide fast lookup. Which EMR big data application can support this technical requirement?

3. Your data scientists are looking for a web-based notebook that they can use for their interactive data analysis and machine learning model development. Your organization has strict security policies that do not allow AWS console access, and also, to support centralized user management, you are looking for a notebook in EMR that can easily support signing in with corporate credentials. Which notebook option in EMR best suits your needs?

Further reading

Here are a few resources you can refer to for further reading:

*   EMR release documentation, which includes EMR releases with big data applications it supports: `https://docs.aws.amazon.com/emr/latest/ReleaseGuide/emr-release-components.html`

*   Detailed steps for setting up EMR Studio: `https://docs.aws.amazon.com/emr/latest/ManagementGuide/emr-studio-set-up.html`

*   IAM permissions and limits while using the AWS Glue Data Catalog as a Hive external Metastore: `https://docs.aws.amazon.com/emr/latest/ReleaseGuide/emr-hive-metastore-glue.html`

Section 2: Configuration, Scaling, Data Security, and Governance

This part of the book will go deep into the advanced configuration of EMR applications, hardware, networking, security, troubleshooting, logging, and the different SDKs/API required to launch and manage EMR clusters. This section will also provide the details of different scaling options and explain the security aspects of EMR such as data protection, authentication, and granular permission management with AWS Lake Formation and Apache Ranger.

This section comprises the following chapters:

- *Chapter 5, Setting Up and Configuring EMR Clusters*
- *Chapter 6, Monitoring, Scaling, and High Availability*
- *Chapter 7, Understanding Security in Amazon EMR*
- *Chapter 8, Understanding Data Governance in Amazon EMR*

5
Setting Up and Configuring EMR Clusters

In previous chapters, while explaining **Amazon EMR** architecture or different big data applications within it, we have given sample AWS CLI commands and a few high-level steps to create an EMR cluster. In this chapter, we will dive deep into setting up an EMR cluster with quick options and also advanced configurations, using which you can control different hardware, software, networking, and security settings.

This chapter will also explain troubleshooting, logging, and tagging features of the EMR cluster and how you can leverage AWS SDKs and APIs to launch or manage clusters.

The following are the topics that we will cover in this chapter:

- Setting up and configuring clusters with the EMR console's quick create option
- Advanced configuration for cluster hardware and software
- Working with AMIs and controlling cluster termination
- Troubleshooting, logging, and tagging a cluster
- SDKs and APIs to launch and manage EMR clusters

Understanding these concepts will help you to have advanced control of your cluster configurations, using which you can reduce maintenance overhead, optimize resources, and also troubleshoot cluster or job failures.

Technical requirements

In this chapter, we will dive deep into an EMR cluster's advanced configuration, logging, and debugging. To test out the configurations, you will need the following resources before you get started:

- An AWS account
- An IAM user that has permission to create EMR clusters, EC2 instances and dependent IAM roles, VPCs, and security groups and can access CloudWatch and CloudTrail logs

Now let's dive deep into an EMR cluster's advanced options and how you can configure them using the EMR console and the AWS CLI.

Setting up and configuring clusters with the EMR console's quick create option

The EMR console's quick create option helps you to create an EMR cluster quickly with default configurations specified for software, hardware, and security sections. Each section has default values selected that you can change or override and there are some configurations that are not exposed for selection during the cluster creation process.

For example, you do not get the option to select a **Virtual Private Cloud** (**VPC**) or subnet for your cluster. EMR configures the cluster in your region's default VPC and public subnet.

Now, to get started, you can follow these steps to create a cluster:

1. After signing in to the AWS console, navigate to the Amazon EMR console at `https://console.aws.amazon.com/elasticmapreduce/`.

2. Choose the **Clusters** option and then select or click **Create cluster**, which will open the **Quick create** page.

3. On the **Create Cluster - Quick Options** page, you will have default values populated that you can change as needed. We will cover the options in detail in a later part of the chapter.

4. Then select **Create cluster**, which will launch the cluster.

5. Then, on the cluster status page, you can find the cluster status, which should change from **Starting** to **Running** and then **Waiting**.

6. When the status reaches the **Waiting** stage, that means it's ready to accept jobs as steps, and also you can SSH to the master node.

Now let's deep dive into the default configuration the quick create page shows, which you might have seen in *step 3*.

It divides the cluster configuration into four sections: **General Configuration**, **Software configuration**, **Hardware configuration**, and **Security and access**. The following screenshot shows options you get in **General Configuration** and **Software configuration**:

Figure 5.1 – General Configuration and Software configuration of the EMR quick creation option

Now let's explore each of the settings you see in **General Configuration** and **Software configuration**:

*   **General Configuration**: This section of the configuration allows you to specify a name for your cluster, enable logging for your cluster, and then define the launch mode, which can be either **Cluster** or **Step execution**.

    *   Logging requires an Amazon S3 path to which EMR will be writing logs. Please note logging can be enabled while creating the cluster only and cannot be changed later. With the quick options, EMR populates a default S3 path that you can override.

- For **Launch mode, Cluster** is selected as the default value that you can change. Cluster mode represents a long-running cluster that does not auto-terminate the cluster when the job executions are completed. If you are creating a transient cluster that needs to execute a few steps and then auto-terminate, then you can select **Step execution** for **Launch mode**.

- **Software configuration**: Amazon EMR has different release versions and it auto-selects the latest version that you can use to create your cluster. Each release includes big data applications and their respective versions included in that release. As a quick option, EMR specifies a set of applications to choose from. For example, **Core Hadoop** includes Hadoop, Hive, Hue, Mahout, Pig, and Tez, and similarly, **HBase**, **Spark**, and **Presto** are also a few other application packages listed that you could choose from. If you need a custom set of applications, then you can use advanced EMR options.

  In addition to selecting an EMR release and big data applications, you will also see the **Use AWS Glue Data Catalog for table metadata option**, which is unchecked by default. This option will enable you to use the AWS Glue Data Catalog as a Hive external metastore.

Now let's understand the configuration settings you get for **Hardware configuration** and **Security and access**.

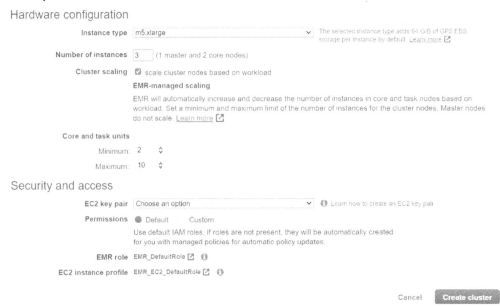

Figure 5.2 – Hardware configuration and Security and access configuration of the EMR quick create option

- **Hardware configuration**: This facilitates choosing the EC2 instance type for your cluster, the number of instances you need, and whether you plan to enable scaling on your cluster.

  By default, EMR specifies **3** as the total instance count for the cluster, which includes 1 master and 2 core nodes. If you would like to change the configuration or would like to configure multiple master nodes, then you can switch to advanced configuration.

  The scaling option is disabled or unchecked by default, but if you plan to enable it, then it requests that you provide the minimum and maximum instance counts for core nodes. By default, it has **2** as the minimum nodes and **10** as the maximum, which you can change as per your requirements. EMR offers two cluster scaling mechanisms, auto scaling and managed scaling, which we will deep dive into in future chapters. With quick option configuration, EMR will use managed scaling to scale the cluster up and down.

- **Security and access**: This section allows you to configure IAM roles, permissions, and an EC2 key pair for your cluster. By default, EMR creates **EMR_DefaultRole** and **EMR_EC2_DefaultRole** with the required permissions, but you can override them with custom roles or permissions you would like to apply.

  EC2 key pair is an optional parameter. You can assign an EC2 key pair if you plan to link to the cluster node using **Secure Shell** (**SSH**) and submit steps or execute commands via the CLI. Please note, you need to create an EC2 key pair first before creating the EMR cluster and it's recommended to assign an EC2 key pair as without it, you won't be able to connect to the cluster master node using SSH.

As we have got an overview of the quick cluster creation option, now let's dive deep into the advanced configurations EMR provides to control cluster hardware, software, networking, and security controls.

Advanced configuration for cluster hardware and software

In the previous section, you saw the default configurations the quick option provides, and now, we will see how you can customize each of those default configurations as per your requirements.

Understanding the Software Configuration section

Using the **Software Configuration** section, you can choose the EMR release, the applications you plan to set up, master node and metastore configurations, and any custom configurations you plan to add that may override default configurations of the cluster.

The following explains each of these configurations.

- **Release**: In this section, you can select the EMR release you plan to use. After selecting the release, you will see the list of applications that release includes with their software version. For each release, EMR automatically marks a few applications as selected, which you are free to change if needed.

- **Multiple master nodes**: EMR supports enabling multiple master nodes, which brings high availability to your cluster. You can select either one or three master nodes. However, launching three master nodes is only supported starting with the EMR 5.23.0 release.

  Higher availability is achieved by using an EMR cluster with three master nodes. If one of the master nodes goes down, then it automatically fails over to another master node without any interruption. Meanwhile, EMR also replaces the failed master node with required configurations and bootstrap actions, so that the cluster maintains three master nodes for high availability. EMR also extends the high availability feature to a few big data applications, such as HDFS, YARN, Hive, Hue, Oozie, and Flink, benefitting from the multiple master feature. For Hive, Hue, and Oozie you need to make sure that the metastore databases are externalized outside of the master node so that the failover to other master node does not affect the operation.

- **AWS Glue Data Catalog settings**: As we have explained in previous chapters, you can externalize your metastore databases outside of your cluster. For that, you can leverage either Amazon RDS, Amazon Aurora databases, or AWS Glue Data Catalog. This setting allows you to specify whether you would like to leverage AWS Glue Data Catalog for Hive and Spark applications. By default, these applications are not selected and you can enable them as needed.

- **Edit software settings**: This section enables you to provide additional custom configuration parameters, which are intended to override the default configurations of the cluster. Configuration objects include a big data application's classification, such as the Hadoop core site, its properties, and other optional nested configurations. You can provide custom configuration as a JSON string or you can save your JSON configuration into Amazon S3 and specify the S3 path while creating the cluster.

For example, the following sample JSON specifies configurations for `core-site` and `mapred-site` classifications and includes Hadoop and MapReduce properties with values that you plan to override in the cluster. In the previous chapter, while covering different big data application configurations, we covered several other examples similar to this.

```
[
    {
        "Classification": "core-site",
        "Properties": {
            "hadoop.security.groups.cache.secs": "500"
        }
    },
    {
        "Classification": "mapred-site",
        "Properties": {
            "mapred.tasktracker.map.tasks.maximum": "10",
            "mapreduce.map.sort.spill.percent": "0.80",
            "mapreduce.tasktracker.reduce.tasks.maximum": "20"
        }
    }
]
```

The following diagram is a screenshot of the AWS console that shows software configurations within the advanced cluster creation options.

Figure 5.3 – Software Configuration of EMR's advanced cluster create option

After understanding what options you have under **Software Configuration**, let's next understand what options you get while configuring steps in your EMR cluster.

Understanding Steps

After specifying software configurations, you can specify steps for your cluster, which may run sequentially or in parallel.

The following are the settings available when you specify optional steps for your cluster:

- **Concurrency**: This setting is disabled or unchecked by default, which means all the steps you add to the cluster will get executed in sequence. By enabling this setting, you can specify a maximum for how many steps or jobs can run in parallel to utilize the cluster resources better.

- **After last step completes**: This setting gives you control to specify whether you would like to terminate your cluster when all defined steps complete execution or you would like to keep the cluster active in the waiting state so that other jobs can be submitted later.

- **Step type**: This setting allows you to add a step to the cluster and the steps can be a Hive, Pig, Spark, or custom JAR file job. Choosing any type opens up an additional configuration screen where you can specify parameters for the step or job.

The following screenshot of the EMR console shows the parameters for steps:

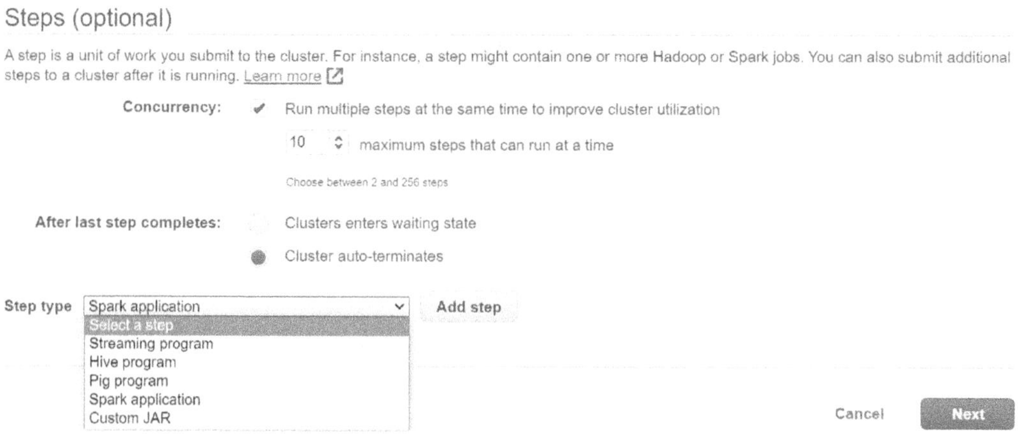

Figure 5.4 – Steps configuration of EMR's advanced cluster create option

In this section, you have learned how you can configure steps in your cluster to get executed in sequence or in parallel. Next, you will learn what hardware configuration options are available.

Understanding the Hardware Configuration section

This section provides configurations using which you can control your cluster hardware, networking, and scaling configurations.

The following explains each of these configurations.

- **Cluster Composition**: This is an important configuration for the cluster hardware where you can specify your cluster nodes to be part of an instance group or instance fleet. An instance group represents a uniform node type for your cluster, whereas an instance fleet represents a mix of different EC2 instance types for core or task nodes. Based on the cluster composition you select, you get additional configuration parameters in cluster node types and cluster scaling configuration.

- **Networking**: This section allows you to launch your cluster in your region's default VPC or other public or private VPC. Apart from configuring a VPC, you can specify one or more subnets for your cluster within which instance nodes will be launched. The subnets can be public, private, or shared, or can be associated with AWS Local Zones or AWS Outposts.

Please make sure you have created the VPC and subnets before so that you can select them while creating your EMR cluster configuration.

The following screenshot shows an EMR console screen that includes **Cluster Composition**:

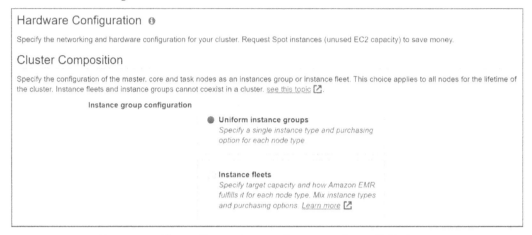

Figure 5.5 – Cluster Composition configuration of your EMR cluster

And here's a screenshot of **Networking** information in the EMR console:

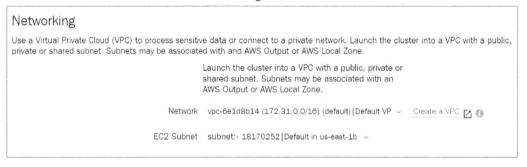

Figure 5.6 – Networking configuration of your EMR cluster

- **Cluster Nodes and Instances**: This section of the configuration provides the control to choose **Instance type**, **Instance count**, and **Purchasing option** options for the **Master**, **Core**, and **Task** node types. The following EMR console screenshot shows where you can select cluster nodes and instances.

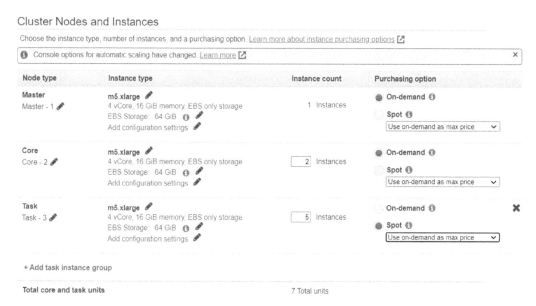

Figure 5.7 – EMR Cluster Nodes and Instances type configuration

Please note, **Task** node types are optional so you can delete them if needed. Also, if you plan to leverage **Task** node types, then you have the flexibility to add multiple task instance groups where you can specify a different instance type and instance count for each task instance group.

- **Cluster scaling**: You can enable scaling capability on your cluster by selecting the **Enable Cluster Scaling** option. It provides the flexibility to choose between EMR managed scaling and a custom scaling policy. With EMR managed scaling you need to provide your minimum and maximum nodes and the maximum number of on-demand and core nodes for scaling.

With a custom auto scaling policy, you can define custom scale-out and scale-in rules with any cluster or application parameters. We will deep dive into this topic in the next chapter.

Please note, when you select **Instance fleets** as the cluster instance group composition, for the scaling feature you only get the EMR managed scaling option as custom auto scaling is not available for instance fleets.

- **EBS Root Volume**: Each instance type you choose will have an EBS root volume attached to it and you can control the size of the volume you need for it. Please note, you can increase the default 10 GB size to a maximum of 100 GB and the same size will be applicable to all instances of the cluster.

The following is a screenshot of the EMR console showing **Cluster scaling** and **EBS Root Volume** size settings:

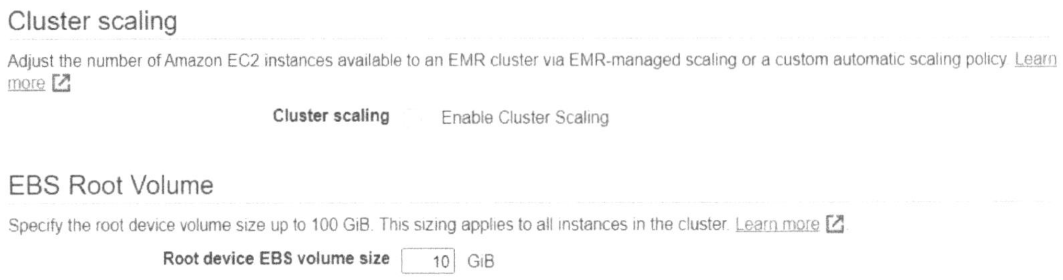

Cluster scaling

Adjust the number of Amazon EC2 instances available to an EMR cluster via EMR-managed scaling or a custom automatic scaling policy. Learn more ↗

Cluster scaling   Enable Cluster Scaling

EBS Root Volume

Specify the root device volume size up to 100 GiB. This sizing applies to all instances in the cluster. Learn more ↗

Root device EBS volume size   10 GiB

Figure 5.8 – EMR Cluster scaling and EBS Root Volume size configuration

In this section, you have learned about configurations related to EC2 instances of the cluster, EBS volumes attached to the EC2 instances, scaling the resources, and also specifying a VPS and security groups. Next, you will learn about a few general configurations that you can apply to your cluster.

Understanding general configurations

This section allows you to configure logging, debugging, tagging, and bootstrap actions on your cluster. This section also provides the option to choose **EMR File System** (**EMRFS**) consistent view and a custom **Amazon Machine Image** (**AMI**) for your cluster.

Now let's get an overview of each of these settings:

- **General Options**: In this section, you can specify your cluster name and specify the configuration for logging, log encryption, debugging, and termination protection. Please note, you can specify these options while launching the cluster only. We will dive deep into each of these sections in future sections of this chapter.

The following screenshot shows the options you get under **General Options**:

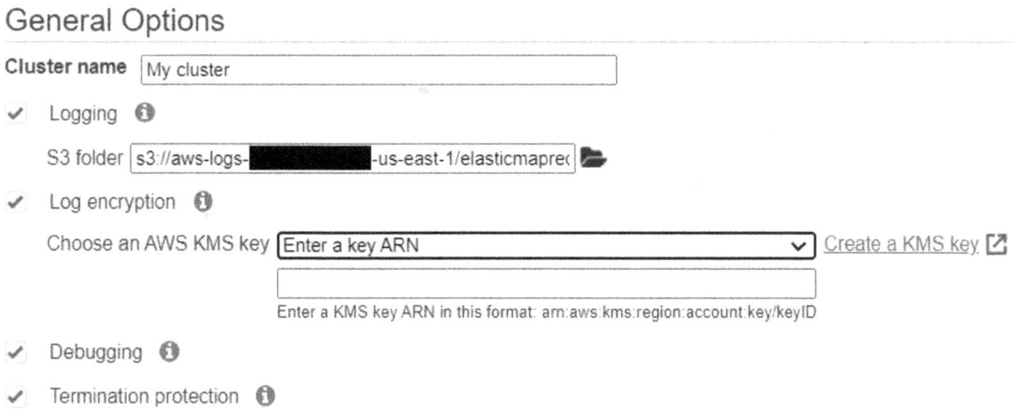

General Options

Cluster name | My cluster

✓ Logging ℹ️

    S3 folder | s3://aws-logs-██████████-us-east-1/elasticmaprec 🗁

✓ Log encryption ℹ️

    Choose an AWS KMS key | Enter a key ARN ⌄ | Create a KMS key ⬀

    Enter a KMS key ARN in this format: arn:aws:kms:region:account:key/keyID

✓ Debugging ℹ️

✓ Termination protection ℹ️

Figure 5.9 – General Options for your EMR cluster

- **Tags**: With tagging, you can group your resources and identify usage and cost by the environment, project, AWS services, and more. It consists of a key and value pair and you can add more than one tag for your cluster. For example, you can add tags to your EMR cluster such as the name of the environment, the project name, project type, the owner, and so on.

  Tags assigned to an EMR cluster are also propagated to its underlying EC2 instances and you can add or remove tags after the cluster is created too. You might also use tags to define IAM permissions.

  The following is a screenshot of the EMR console that represents assigning multiple tags to your cluster:

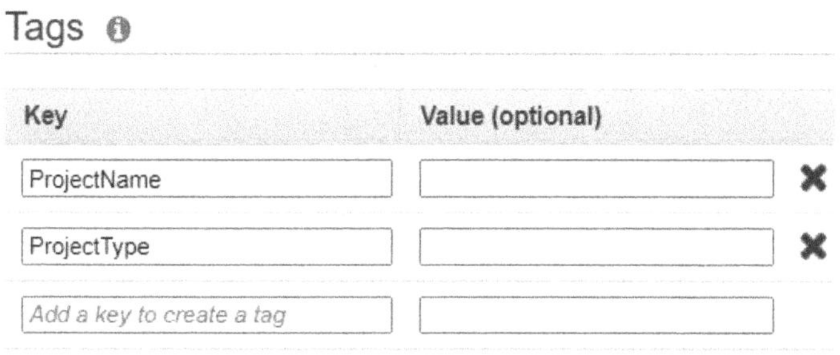

Tags ℹ️

Key	Value (optional)	
ProjectName		✖
ProjectType		✖
Add a key to create a tag		

Figure 5.10 – Adding tags for your EMR cluster

- **Additional Options**: This section allows you to specify a custom AMI ID that you plan to use for your cluster nodes and any bootstrap actions you would like to add. By default, your EMR cluster uses Amazon Linux AMI for Amazon EMR but starting from the EMR 5.7.0 release, you can override it by providing your custom AMI.

  If you would like to install custom libraries and software on your cluster nodes, then you have two options to do it. Either you can specify bootstrap actions with the command you would like to execute or create a custom AMI with all the required software and use that to launch your cluster. Please note a custom AMI performs better compared to bootstrap actions as bootstrap actions are executed after the cluster is created.

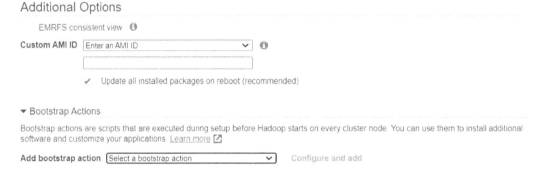

Figure 5.11 – Additional Options for your EMR cluster

> **Important Note**
>
> EMR provides the **EMRFS consistent view** option to solve the eventual consistency issue of Amazon S3. But recently Amazon S3 started supporting strong read-after-write consistency, which means you won't face the eventual consistency issue.
>
> You don't need to enable the **EMRFS consistent view** option for any of the EMR release versions.

In this section, you have learned about general configurations using which you can specify logging, debugging, and tagging options for your cluster and also can specify custom AMIs with bootstrap actions. In the next section, you will get an overview of the options EMR has related to the security of the cluster.

Understanding Security Options

The **Security Options** section gives you configurations to specify EC2 key pairs, IAM roles, security groups, and more, using which you can control who can access your cluster resources and what privileges they have. By default, EMR creates roles and access policies as needed by the cluster, but you can override them with your custom roles and access policies.

Now let's get an overview of each of these settings:

- **EC2 key pair**: As explained regarding the quick create option, EC2 key pairs are used to link to the master node using SSH and it is recommended to assign an EC2 key pair while creating the cluster as you cannot assign one later.

- **Permissions**: By default, EMR creates three roles – `EMR_DefaultRole` as the EMR role, `EMR_EC2_DefaultRole` as the EC2 instance profile, and `EMR_AutoScaling_DefaultRole` as the Auto Scaling role. But you can create your own custom roles and assign them to the cluster.

  To understand each of these roles better, EMR role calls or interacts with other AWS services, such as EC2, while creating the cluster. The EC2 instance profile role provides access to cluster EC2 instances to access other AWS services such as Amazon DynamoDB, Amazon S3, and more. The Auto Scaling role provides access to add or remove EC2 instances from the cluster when scaling up or down happens through managed scaling or auto scaling policies.

- **Security Configuration**: This allows you to specify encryption and authentication options for your cluster. You need to create the configuration before creating the EMR cluster.

- **EC2 security groups**: EC2 security groups provide firewall security for your AWS services' inbound and outbound access. You can control which ports are allowed to access which source IP or security group or whether it is open to all.

  EMR establishes two security groups by default: one for the master node and another for the core and task nodes. You can choose **EMR managed security groups**, which EMR automatically updates as needed or you can create your custom security groups to control access to your cluster.

The following is a screenshot of the EMR console that includes the security options you get when you create an EMR cluster with advanced options.

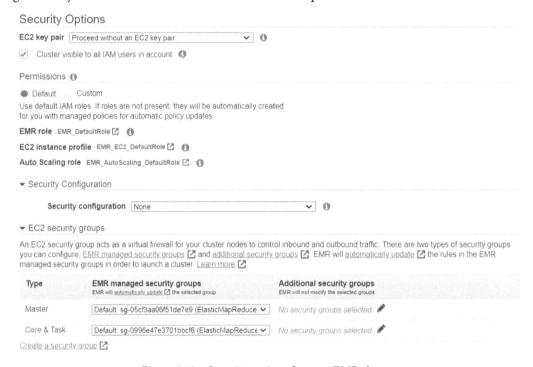

Figure 5.12 – Security options for your EMR cluster

After selecting options on the final security screen, click **Create cluster**, which will launch the cluster and start executing steps if defined or will go into the waiting state.

In this section, you have learned how you can use the EMR console's advanced options to create your EMR cluster. In the next section, let's dive deep into working with custom AMIs and how you can control cluster termination.

Working with AMIs and controlling cluster termination

In the previous section, we explained how EMR by default uses the Amazon Linux AMI for EMR and you have the option to create a custom AMI and use it while creating a cluster.

Now, in this section, we will dive deep into the default Amazon Linux AMI for EMR, custom AMI implementations, and how cluster termination works that you can configure as per your use case.

Working with AMIs

An AMI includes all the resources required to launch an EC2 instance. While launching an instance, you can specify the AMI it should be using. You can use the same AMI to launch multiple EC2 instances. If your EC2 instances need different configurations, then you can create instance-specific AMIs.

An AMI has the following components:

- One or more **Elastic Block Store** (**EBS**) snapshots, or if the AMI is backed by an instance store, a template for the EC2 instance's root volume. This might include an operating system, applications, and application servers.

- Permissions that defines which AWS account can use this AMI to launch EC2 instances.

- A block device mapping, which defines the volumes to be attached when an instance is launched.

Next, we will learn how EMR uses the default AMIs available and what configurations are available to specify your custom AMIs.

Using the default Amazon Linux AMI for EMR

For each EMR release version, there is a predefined and pre-configured AMI available that is integrated with the big data applications in that release. This means even if a new Amazon Linux AMI is available for that release, it won't be used in EMR. This is the reason it is recommended to use the latest release of EMR unless you have a specific need to use earlier releases of EMR.

The following explains how software updates are handled in the default AMI:

- When the EMR cluster's EC2 instance boots for the first time, it uses the default Amazon Linux AMI for Amazon EMR, identifies the package repository enabled for the AMI, and checks for software updates that apply to the AMI version. Similar to other EC2 instances, important security updates are applied automatically from these repositories.

  Please note your networking and firewall settings should allow egress traffic to the Amazon Linux repository in S3.

- By default, software packages or kernel updates that require instance reboot are not automatically downloaded while launching the instance.

- The cluster completes the launch irrespective of whether the package installation is successful or not. If for any reason there is a network issue or the repository is not reachable and the software packages cannot be installed, that does not stop the cluster launch. So provide additional monitoring to keep a check on the cluster.

> **Important Note**
>
> It is recommended not to run `sudo yum update` on cluster EC2 instances either through the SSH command line or using bootstrap actions. This might create incompatibilities between nodes and big data applications.

The following are a few considerations or best practices that you can follow while using the default AMI:

- If you are using an earlier release of EMR, before updating software packages, consider testing the migration with the latest release.

- If you plan to migrate to the latest release, then test the implementation in a non-production environment. To do this, you can leverage the cloning feature for the EMR cluster.

- Look at Amazon Linux Security Center for any updates.

- Avoid installing custom packages by directly doing SSH to individual cluster nodes as it might create inconsistencies across nodes. Instead, use bootstrap actions for additional custom installations, which will install required software packages across the cluster. This requires terminating the cluster and relaunching it.

Using a custom AMI with your EMR cluster

In EMR, custom AMIs are supported starting with the EMR 5.7.0 release. It's useful to use a custom AMI when you need to do the following:

- Perform software customizations or pre-install applications before using the cluster. As explained earlier, you can do customizations using bootstrap actions too but pre-installed or pre-configured AMIs help in reducing the time required to launch an EMR cluster as the customizations are already a part of the cluster and no additional step is needed. This way, custom AMIs can improve your cluster start time.

- Implement more automated, sophisticated node and cluster configurations than bootstrap action steps allow.

Earlier than the EMR 5.24.0 release, you could use encrypted EBS root volumes only if you are using custom AMIs. But after EMR 5.24.0, you have the option to specify encryption using EMR security configurations. Having gotten an overview of default and custom AMIs, next we will learn the difference between cluster-level and instance-level AMIs.

Cluster-level versus instance-level custom AMIs

While assigning custom AMIs, you have the option to assign a single custom AMI to the whole cluster or an instance-level custom AMI, which will have a different custom AMI for each instance of your cluster.

> **Important Note**
> Please note, you cannot have both instance-level and cluster-level AMIs assigned at the same time.

Starting from EMR release 5.7.0 and later, you have the option to specify a different instance-level custom AMI for each instance type in an instance group or instance fleet. For example, you can configure arm64 architecture that is available with m6g.xlarge instance types and x86_64 architecture that is available in m5.xlarge instance types in the same instance fleet or group. Each instance type will use a custom AMI that matches its application architecture.

The following are some differences you will find when comparing cluster-level custom AMIs with instance-level AMIs:

- Instance-level custom AMIs can use both Graviton based ARM and x86 architectures in the same cluster, which you cannot do with single cluster-level AMIs.

- You can assign a custom AMI when adding an instance fleet or instance group to an already running cluster only with instance-level AMIs.

- Be it cluster-level or instance-level custom AMIs, you cannot run a custom AMI and an EMR AMI in the same cluster.

The following are a few best considerations or best practices that you can follow when creating your EMR cluster with custom AMIs:

- EMR 5.30.0 and later and the Amazon EMR 6.x series clusters are based on Amazon Linux 2, so for them, you need to use custom AMIs based on Amazon Linux 2. For EMR releases earlier than 5.30.0, you should use a 64-bit Amazon Linux AMI.

- Use the most recent EBS-backed Amazon Linux AMI as the base for your customizations.

- Avoid copying a snapshot of an existing cluster EC2 instance to create a custom AMI as that causes errors.

- Make sure you choose only the HVM virtualization type and instances that are compatible with Amazon EMR. Please check the AWS documentation to find the supported instance types.

- Your cluster service role should have launch permissions on the AMI or your AMI should be public to grant access, or you might be the owner of the AMI or the owner of the AMI shared it with you.

- Avoid using application names such as "Hadoop," "Hive," or "Spark" as user names for the AMI, as they might conflict with cluster service names and cause errors.

- If your AMI has contents in the /var, /tmp, and /emr folders, then they are moved to the respective /mnt/ path of the cluster (for example, /mnt/var, / mnt/tmp, and /mnt/emr). Please note, these file contents are reserved so if you have a large amount of data in these paths, then that might affect your cluster startup time.

For more detailed considerations and limitations, please refer to the AWS documentation, which might include more updates with new EMR releases.

> **Important Note**
> Please note, while using a custom AMI for your EMR cluster, use the **Update all installed packages on reboot** option, which is recommended.

In this section, you got an overview of default and custom AMIs in EMR and how you can configure them at the instance or cluster level. In the next section, we will dive deep into the EMR cluster termination process and how you can control it.

Controlling the EMR cluster termination process

As explained in the previous section, while creating an EMR cluster you can configure it to be long-running or you can configure it to auto-terminate after the defined steps have completed successful execution. When a cluster gets terminated, all its associated Amazon EC2 instances are terminated and you lose access to data in the EBS volume or instance store. The content of the EBS volume or instance store is not recoverable, which means you should have a good strategy in place for terminating your cluster.

Auto-terminating the cluster in transient cluster use cases optimizes your costs as the cluster does not spend time in the waiting state and you pay for actual usage hours. In long-running clusters that require interactive analytics, you can enable termination protection to prevent accidental termination of the cluster.

There are a few important points that you should consider:

- EMR clusters created through the AWS console or the AWS CLI do not have auto-termination enabled by default, whereas clusters created through the EMR API have auto-termination enabled by default.

- EMR enables auto-termination for clusters that have multiple masters and it overrides any settings you provide for auto-termination.

- To terminate a cluster with multiple masters, you need to disable the auto-termination settings first and only then will it allow you to terminate the cluster.

Now let's get more insights into how you can leverage auto-termination and termination protection features of EMR.

Configuring an EMR cluster to auto terminate

If you plan to auto terminate your cluster after your steps are done, then you can look at the following options to configure auto-termination:

- **Using the quick creation option of the AWS console**: When you create a cluster using the quick create option of the AWS console, auto-termination is only enabled if you follow the **Step execution** launch mode.

- **Using the advanced option of the AWS console**: When you create a cluster using the advanced option, under the **Steps (optional)** configuration, select **Cluster auto-terminates** for the **After last step completes** setting.

- **Using the AWS CLI**: When creating a cluster with AWS CLI commands, you can enable auto-termination for your cluster using the - -auto-terminate option. The following is a sample AWS CLI command that includes auto-termination configuration:

```
aws emr create-cluster --name "EMR Cluster"
--release-label emr-6.3.0 --applications Name=Hive
Name=Spark --use-default-roles --ec2-attributes
KeyName=myKey --steps Type=HIVE,Name="Hive Program",
ActionOnFailure=CONTINUE, Args=[-f,s3://<mybucket>/
scripts/query.hql,-p,INPUT=s3://<mybucket>/
inputdata/,-p,OUTPUT=s3://<mybucket>/
outputdata/,$INPUT=s3://<mybucket>/
```

```
inputdata/,$OUTPUT=s3://<mybucket>/outputdata/]
--instance-type m5.xlarge --instance-count 4 --auto-
terminate
```

This section explained how you can configure the auto-termination of your cluster when you have transient EMR cluster use cases. Now let's look at long-running cluster use cases, where you can enable termination protection to prevent accidental termination.

Using termination protection for your long-running cluster

When you have enabled termination protection on your long-running cluster, you can still terminate it but with an additional step where you have explicitly disabled the termination protection first and then terminate your cluster. This feature helps you prevent terminating the cluster because of any error.

When termination protection is enabled on your cluster, the `TerminateJobFlows` action of the EMR API fails so you cannot terminate the cluster using the API or AWS CLI. The AWS CLI exits the command execution with a non-zero return code and the EMR API returns an error but if you are using the EMR console, then you will be prompted if you need to turn off the termination protection.

> **Important Note**
>
> Please note that termination protection prohibits you from terminating the cluster but does not provide any guarantee against data loss or rebooting the EC2 instances that might be caused because of human error. You can still trigger an EC2 instance reboot when you are connected to any instance through SSH or you have an automated script that triggered a reboot of the instance.
>
> With termination protection, there are chances your HDFS data will still be lost. Leverage Amazon S3 as your persistent data store to prevent data loss because of EC2 instance failures.
>
> Also note that termination protection does not affect your cluster when you resize your cluster with auto scaling policies or when you add or remove instances from your cluster's instance group or instance fleet.

Next, we will learn how EMR termination protection works when your cluster uses EC2 Spot instances.

Termination protection with EC2 and Spot instances

When you enable termination protection for your EMR cluster, it has the `DisableAPITermination` attribute set for all the Amazon EC2 instances of the cluster. Please note, you do have separate termination protection configuration available for your EC2 instances and when you trigger a termination request for your EMR cluster and the settings for EMR and EC2 instances conflict, then EMR overrides the EC2 instance settings.

Let's assume you have used the EC2 console to enable termination protection for your EC2 instances but your EMR cluster has termination protection disabled. In that state, if you terminate your EMR cluster, then it will set `DisableApiTermination` to false on the associated EC2 instances and then terminate the instance and the cluster.

> **Important Note**
>
> Please note that the termination protection setting of your EMR cluster does not apply to EC2 Spot instances. If you are using EC2 Spot instances for your cluster's core or task nodes and the Spot price rises beyond the maximum Spot price you have defined, then Spot instances will get terminated irrespective of the termination protection setting on the cluster. With **Use on-demand as max price**, you can avoid Spot instance termination.

Having learned how termination protection works with EC2 and Spot instances, next we will learn how termination protection works when we have unhealthy YARN nodes.

Termination protection behavior with unhealthy YARN nodes

As we explained in *Chapter 2*, *Exploring the Architecture and Deployment Options*, EMR periodically checks the status of YARN applications running on instances and also the instance's health by checking the status of NodeManager's health checker service. If any specific node is reported as **UNHEALTHY**, then the EMR instance controller does not allocate any new containers to it and blacklists the node until it becomes healthy again.

There can be multiple reasons for an EC2 instance becoming unhealthy, but a common reason is the instance disk utilization goes beyond 90%. If a node continues to be **UNHEALTHY** for more than 45 minutes, then Amazon EMR takes the following actions, depending on whether the cluster has termination protection enabled or not:

- If termination protection is enabled, the unhealthy EC2 core instances continue to remain in the blacklisted state and will be counted towards your cluster capacity or cost. As the EC2 instances are not terminated, you can connect to your EC2 instance, make configuration changes, or perform data recovery and can resize your cluster to add additional node capacity.

However, unhealthy task nodes are not protected from termination and get terminated if they continue to stay unhealthy for more than 45 minutes.

- If termination protection is disabled on your cluster, then EMR terminates the EC2 instances irrespective of task or core nodes, but to maintain the minimum capacity specified in the instance group or instance fleet, it provisions new instances.

If all of the core nodes of the cluster are reported as **UNHEALTHY** for more than 45 minutes, the complete EMR cluster will be get terminated with the **NO_SLAVES_LEFT** status.

When the instances get terminated, the HDFS data will get lost and you will not have a way to recover them. So it is recommended to enable termination protection on your cluster and also use Amazon S3 as a persistent data store instead of instance EBS volumes as an HDFS store.

Termination protection, auto-termination, and step execution

When you have enabled both termination protection and auto-terminate settings with step execution, then auto-termination takes precedence, which terminates the cluster after finishing all step execution.

When you define steps on your cluster, you do have the option to configure what action should be taken when a step fails. You can set the `ActionOnFailure` property value to define the action on step failure, which has values of `CONTINUE`, `CANCEL_AND_WAIT`, and `TERMINATE_CLUSTER`.

If auto-termination is enabled and `ActionOnFailure` is set to `CANCEL_AND_WAIT`, then the cluster gets terminated without executing any other subsequent steps.

If `ActionOnFailure` is set to `TERMINATE_CLUSTER`, then the cluster terminates in all cases except when auto-termination is disabled and termination protection is enabled.

Configuring termination protection while launching your cluster

You can enable termination protection while creating your cluster using the AWS console, the AWS CLI, or using the EMR API. Termination protection is disabled by default in all the approaches except when you create a cluster using the AWS console's advanced options.

Now let's learn how you can configure termination protection for your cluster while creating it through the AWS CLI or AWS console:

- **Using the AWS Console**: When you create a cluster using advanced options, go to **General Cluster Settings** under **General Options** and select the **Termination protection** option to enable it or uncheck it to disable it. The following screenshot represents the option in the AWS console.

Figure 5.13 – Termination protection configuration with EMR advanced cluster create options

- **Using the AWS CLI**: When creating a cluster with the AWS CLI commands, you can enable termination protection on your cluster using the `--termination-protected` parameter. The following is a sample AWS CLI command that includes termination-protected configuration:

```
aws emr create-cluster --name "EMR Cluster"
--release-label emr-6.3.0 --applications Name=Hive
Name=Spark --use-default-roles --ec2-attributes
KeyName=myKey --steps Type=HIVE,Name="Hive Program",
ActionOnFailure=CONTINUE, Args=[-f,s3://<mybucket>/
scripts/query.hql,-p,INPUT=s3://<mybucket>/
inputdata/,-p,OUTPUT=s3://<mybucket>/
outputdata/,$INPUT=s3://<mybucket>/
inputdata/,$OUTPUT=s3://<mybucket>/outputdata/]
--instance-type m5.xlarge --instance-count 4
--termination-protected
```

Now let's learn about termination protection.

Configuring termination protection for a running cluster

In the previous section, you learned how you can configure termination protection while launching or creating a cluster, but you also have the option to change the settings for an already running cluster.

You can change the setting through both the AWS console and the AWS CLI:

- **Using the AWS console**: To change the termination protection configuration for a running cluster, you can navigate to the EMR console, select the cluster for which you plan to change the configuration, and then on the **Summary** tab, for **Termination protection**, choose **Change**, which will provide the following options:

Figure 5.14 – Change the Termination protection configuration

Select **On** or **Off** and select the green check mark to confirm it.

- **Using the AWS CLI**: To change the termination protection configuration for a running cluster using the AWS CLI, leverage the `modify-cluster-attributes` EMR CLI command with the `–termination-protected` or `–no-termination-protected` parameters. The following is a sample AWS CLI command to enable termination protection for a running cluster:

```
aws emr modify-cluster-attributes --cluster-id
<cluster-id> --termination-protected
```

The following is a sample AWS CLI command to disable termination protection for a cluster if it is already enabled:

```
aws emr modify-cluster-attributes --cluster-id
<cluster-id> --no-termination-protected
```

Before executing the preceding commands, please replace `<cluster-id>` with your EMR cluster ID.

In this section, you have learned about an EMR cluster's default AMI and how you can configure a custom AMI. You have also learned about the cluster termination process.

In the next section, you will get an overview of how you can troubleshoot your cluster failures and what logging options you have that can help troubleshoot your cluster.

Troubleshooting and logging in your EMR cluster

An Amazon EMR cluster has several components, such as open source software, custom application code, and AWS integrations, which can contribute to cluster failures or can take longer than expected to complete defined jobs. In this section, you will learn how you can troubleshoot these failures and what fixes can be applied.

When you are starting to implement big data applications in an EMR cluster, it's recommended to enable debugging on the cluster and also take a step-by-step approach to test your application with a smaller subset of data, which might help in debugging failures.

Let's dive deep into a few troubleshooting aspects that can help.

Tools available to debug your EMR cluster

We can divide the set of tools available for troubleshooting into the following three categories:

- Tools that display cluster details
- Tools to view cluster or application logs
- Tools that can be used to monitor cluster performance

Now let's dive deep into each of these sections.

Tools that display cluster details

You can leverage the AWS EMR console, AWS CLI commands, or EMR APIs to get cluster details or any specific job details:

- **Using the AWS console**: On the EMR console, you can see a list of active or terminated clusters that you have launched in the past 2 months. You can select the cluster name for which you would like to get more details and the cluster detail screen provides information with a multiple-tab structure that includes a summary, application user interfaces, monitoring, hardware, and more.

  The **Application user interface** tab of the console provides more details about YARN or other applications' status, such as Spark, where you can drill down to find different metrics, job stages, and executors assigned to them. This interface is available for EMR clusters with a release version of 5.8.0 or more.

- **Using the AWS CLI**: You can get cluster details using the AWS CLI command by passing the –describe parameter.

- **Using the EMR API**: You can leverage DescribeJobFlows of the EMR API to get details about a specific cluster.

Having learned how you can find cluster details; next we will look at what tools we have to view log files.

Tools to view log files

Both Amazon EMR and big data applications on the cluster generated different log files and you can access these log files which depends on the configuration that you specified while creating the cluster.

The following are some of the ways you can access logs:

- **Log files on the cluster master node**: Every cluster publishes its logs to the /mnt/var/log/ path of the master node, which is accessible till the time the cluster is active.

- **Log files archived in Amazon S3**: While launching the cluster, if you have specified an Amazon S3 path, then EMR copies the master node logs available in /mnt/var/log/ to S3 every 5 minutes. This helps you persist the log files, which you can access after the cluster is terminated too. As the log files are copied every 5 minutes, a few last-minute logs might not be available when the cluster is being terminated.

Having understood how you can access the log files of your cluster, in the next section, you will learn how you can monitor your cluster's performance.

Introducing tools to monitor cluster performance

To monitor your cluster usage and performance, you have primarily two options. One is Hadoop application web interfaces that you can access to monitor respective big data applications and the other is Amazon CloudWatch, which can be used for centralized logging too:

- **Hadoop application web interfaces**: Depending on the big data applications you have configured on your cluster; you can access web interfaces available for them using an SSH tunnel through the cluster master node. You can learn about this a bit more in the next chapter.

- **Amazon CloudWatch metrics**: EMR clusters publish various metrics to CloudWatch, which you can use for monitoring or defining alarms with CloudWatch rules.

In this section, you have learned about different tools available in your EMR cluster for viewing cluster details, accessing logs, and monitoring applications. In the next section, you will learn how you can view and restart different EMR applications.

Viewing and restarting cluster application processes

While troubleshooting or monitoring your cluster, you might be interested to list the application processes running in your cluster and for any configuration changes, you might need to restart them.

There are two types of processes that run on a cluster. One is EMR processes, which can be instance-controller or LogPusher, and the other is related to your Hadoop application-related processes, for example, `Hadoop-yarn-resourcemanager` or `Hadoop-hdfs-namenode`.

Now let's get an overview of how you can view or restart these application processes.

Viewing running processes of your cluster

To view the list of Amazon EMR processes, you can execute the following command on your cluster master node's Linux prompt:

```
ls /etc/init.d/
```

This command will provide output as follows:

```
acpid         cloud-init-local        instance-controller        ntpd
```

To view the list of processes related to the application released, you can execute the following command on your master node's Linux prompt:

```
ls /etc/init/
```

This command will provide output as follows:

```
control-alt-delete.conf          hadoop-yarn-resourcemanager.conf
hive-metastore.conf
```

In this section, you have learned about identifying running processes and in the next section, you will learn how you can restart them.

Restarting processes

After you identify the processes running, you might need to stop, start, or restart them. Depending on whether it's an Amazon EMR process or Hadoop application process, you will have a different command to restart the processes.

To stop, start, or restart Amazon EMR processes, you can execute the following commands:

```
sudo /sbin/stop <process-name>
sudo /sbin/start <process-name>
```

To restart the processes related to EMR application releases, you can execute the following commands:

```
sudo /etc/init.d/<process-name> stop
sudo /etc/init.d/<process-name> start
```

Please replace `<process-name>` in the preceding commands with the actual process you plan to stop and start.

Troubleshooting a failed cluster

This section will explain how you can troubleshoot a cluster that has failed, which means it is terminated with an error code. It will cover the following steps:

- Step 1: Collecting data about the issue
- Step 2: Checking the environment
- Step 3: Checking the last state change
- Step 4: Looking at the log files
- Step 5: Testing the cluster step by step

Now let's dive deep into each of these steps.

Step 1: Collecting data about the issue

As a first step, you need to gather information about your cluster that includes collecting details on the issue, cluster configuration, and status.

- **Define the problem**: When you start investigating the issue, you can collect details by asking a few high-level questions such as *What was expected to happen and what really happened? When was the first occurrence of the issue? How frequently or how many times has the issue occurred? Did we change anything in the cluster configuration that was not planned for?* And so on. The answers to these questions will provide a great starting point to troubleshoot the issue.

- **Collect cluster details**: Collect your cluster details, which include the cluster identifier, the AWS region, availability zones, the number of masters, core and task nodes, types of EC2 instances, and whether you configured an instance group or instance fleet, which might help in identifying whether there are limitations around the maximum number of instances you can provision.

After you have collected these basic details, next you can check the environment.

Step 2: Checking the environment

As the next step, you can check for any service outages or usage limits that caused the failure, or the issue could be specific to your EMR release version or related to networking configurations:

- **Check for service outage**: When you create a cluster, under the hood, EMR uses several AWS services, including Amazon EC2 instances for cluster nodes, Amazon S3 to store logs or cluster data, CloudWatch for log monitoring, and many more. The failure could be related to any of these services, so checking the status of the services will help, which is accessible through `https://status.aws.amazon.com/`.

- **Check usage limits**: Every AWS service has a default quota limit set, which can be increased upon request. When you are creating a cluster, it might hit any specific service limits, which could be the number of Amazon EC2 instances launched in your region or it could be the number of S3 buckets you can create. You can check for the error message; for example, if you are hitting the EC2 quota limit for your account, then you might get an **EC2 QUOTA EXCEEDED** error.

- **Check the EMR release version**: Check the EMR release you selected while launching the cluster. As the cluster includes several pieces of open source software, the issue you are facing might have been fixed in the latest EMR releases. In those cases, you can re-launch your cluster with the latest EMR release.

- **Check the cluster VPC and subnet configuration**: Please check if you have configured your VPC or subnet settings as described in the AWS documentation and also make sure that your subnet has enough IP addresses to assign to cluster nodes.

As the next step, we can look at the latest state change of your cluster.

Step 3: Checking the last state change

Look at your cluster's last state change, which might provide information about what happened when it changed status to **FAILED**. For example, when you launch a cluster with a Spark Streaming step and your output S3 path already exists, then you get the error **Streaming output directory already exists**.

You can get the last state change with the AWS CLI `describe-cluster` and `list-steps` commands, or with the EMR API's **DescribeCluster** and **ListSteps** actions.

As the next step, we can look at the log files for further debugging.

Step 4: Looking at the log files

The next step is to examine the logs being generated by your cluster, which can be instance `syslogs` or different Hadoop application logs. If the initial task attempt does not complete on time, EMR might terminate it and create a duplicate task attempt, which is called a speculative task. This activity will generate a significant amount of logs, which get logged into `stderr` or `syslog` of the instances.

For your debugging, you can start checking bootstrap action logs for any unexpected configuration changes or for any errors. Then you can look at each step log to identify whether there were any errors in any step that caused the failure. You can also look into Hadoop job logs to discover failed task attempts.

Now let's get an overview of each of these log types:

- **Check bootstrap action logs**: Bootstrap actions are intended to run startup scripts on your cluster as it is launched. Their primary purpose is to install additional software libraries or customize default configurations. There is a chance that these bootstrap scripts created the failure or affected the cluster's performance, so checking its logs will provide additional insights.

- **Check your cluster step logs**: There are four types of logs generated from each step of the cluster: `controller`, `stderr`, `stdout`, and `syslogs`.

  `controller` logs contain errors generated by Amazon EMR while trying to execute your step. Errors generated from accessing your application steps or loading are often included here. `syslog` primarily includes non-Amazon software logs, which might point to open source Apache Hadoop or Spark streaming errors.

  `stdout` logs include the status of mapper and reducer task executables. Often, application loading errors are included here and sometimes contain application error messages too. `stderr` includes error messages that are generated while processing your defined steps. This log sometimes may contain stack traces or application loading errors.

  For any obvious errors, `stderr` logs are very helpful. They could provide a list of errors if the step got terminated quickly by throwing errors that might be related to mapper or reducer applications running on the cluster.

  You can also check the last few lines of your controller or `syslog` as that might include notices of failures or errors related to failed tasks, if it says **Job Failed**.

- **Check task attempt logs**: If you notice one or more failed tasks from your previous analysis, then analyzing the task attempt logs might provide more insights too.

After analyzing all the logs, you should plan for step-by-step testing of your cluster, which might help debug the issue.

Step 5: Testing the cluster step by step

Restarting your cluster without any steps and adding steps one by one to debug is a great technique that might help too. This way, you can see the failures of any step and you can try to fix and rerun to validate.

The following is an approach you can follow for your step-by-step execution:

1. Launch a new cluster with the previous configuration and two additional configurations (if not enabled earlier), that is, to keep the termination protection and keep alive enabled. This will help follow a step-by-step approach and prevent creating a new cluster every time.

2. Once your cluster is in the **WAITING** state, you can submit your steps one by one.

3. When your step completes processing, look for errors in that specific step's log files. You can connect to the master node with SSH to view the logs. Please note, step log files takes some time to appear.

4. If the step succeeded without any errors, then you can run the next step. If not, then analyze the logs to find the error. If it is an application code error, then apply the necessary fix and rerun this step.

5. Once your debugging steps are done, you can disable the termination protection and then terminate the cluster.

In this section, we have provided a few steps using which you can troubleshoot a failed cluster. Next, we'll see how you can troubleshoot a cluster that is running slowly.

Troubleshooting a slow cluster

This section will explain how you can troubleshoot a cluster that is in the running state but takes longer than expected to return results. Most of the time, it might be caused by resource constraints for your job and might get resolved by assigning more resources, either by moving to high instance types or by increasing the number of instances.

Apart from resource constraints, there might be other reasons that are making your jobs run slowly and the following steps might help in identifying them:

-   Step 1: Collecting data about the issue

-   Step 2: Checking the environment

-   Step 3: Looking at the log files

-   Step 4: Checking your cluster and instance health

-   Step 5: Looking for suspended instance groups

-   Step 6: Reviewing cluster configuration settings

-   Step 7: Validating your input data

Now let's dive deep into each of these steps.

Step 1: Collecting data about the issue

Similar to the failed cluster scenario, step 1 should be asking high-level questions about expectation versus reality, configuration changes, and the frequency of errors, and then collecting cluster details including availability zones, region, VPC, subnets, EC2 instance type configurations, and more.

Step 2: Checking the environment

Checking the environment step is also the same as in the failed cluster scenario, where you check for any service outages, usage limits, and networking configurations that might affect the cluster's expected behavior.

Sometimes environment issues might be transient and restarting the cluster might help in improving the performance.

Step 3: Looking at the log files

As explained for the failed cluster scenario, looking at bootstrap action logs, step logs, and task attempt logs provides a great level of detail about the failure of a step or slow-running jobs.

In addition, checking the Hadoop daemon logs also helps us, which are available in `var/log/Hadoop` of each node. You can also look for failed task nodes or instances from the JobTracker logs and then connect to that instance to find any instance-specific issues related to CPU or memory usage.

Step 4: Checking your cluster and instance health

As you learned earlier, your EMR cluster consists of three types of nodes that include master nodes, core nodes, and task nodes. Each of these node types might contribute to slow-running jobs as they go through resource constraints such as CPU and memory or experience network connectivity issues.

When you are looking at your cluster health, you should look at both cluster- and individual instance-level health. There are several tools that you can use for monitoring health and the following are some of the commonly used methods:

- **Check for service outage**: EMR clusters push different metrics to Amazon CloudWatch, including the performance of the cluster, HDFS utilization, total load, running or remaining tasks, and more. You can leverage these metrics to get an overall picture of your cluster and jobs and also can define alarms to get notified if any metrics go beyond a threshold.

- **Check job status and HDFS health**: On the EMR console, on the **Application user interfaces** tab, you can look at YARN application details and can drill down to logs for checking the status of jobs.

A few Hadoop or big data applications have web user interfaces for monitoring tasks such as JobTracker, HDFS NameNode, TaskTracker, or Spark HistoryServer that you can leverage to identify the amount of resources being consumed by each task or Spark executor, which node they are running, and whether there are resource constraints that you should be working to resolve.

- **Check EC2 instance health**: You should also look at individual EC2 instance health in the EC2 console and can also define CloudWatch alarms for monitoring and notifications.

After looking at these, next, you should look at your instance groups if they are in the suspended state.

Step 5: Looking for suspended instance groups

As discussed earlier, you can define instance groups while configuring your cluster and there is a chance the instance group itself might go into the **SUSPENDED** state if it continues to fail new nodes or check in with existing nodes.

The launch of a new instance or node might fail if Hadoop or related services are broken in some way and do not accept new nodes, or there is a bootstrap script configured for new nodes that fails to complete, or the node itself is not working as expected and is not able to check in with Hadoop. If the issue persists for some time, instead of provisioning new nodes, the instance group goes into the **SUSPENDED** state.

If the instance group goes into the **SUSPENDED** state and the cluster is in the **WAITING** state, then you can add a cluster step to reset the required number of core or task nodes, which might resume the instance group back to the **RUNNING** state.

Step 6: Reviewing cluster configuration settings

When you launch a cluster, Amazon EMR uses default Hadoop configurations, which you can override using bootstrap actions. These configuration parameters are used to execute your jobs and the job log data is stored in a file called `job_<job-id>_conf.xml`, which is stored in the `/mnt/var/log/hadoop/history/` directory of the cluster's master node.

You can review jobs and override the default configuration parameters as needed to improve your job's performance.

Step 7: Validating your input data

One other thing you can check is your input data quality and distribution across nodes or executors. There is a chance that your data is not evenly distributed, which means a single node or Spark executor might be overloaded with a big chunk of your data. This uneven distribution of data is represented as data skewness, which results in one node or Spark executor getting stuck for a long period of time as it needs to process most of the data.

You can also look at data quality as corrupted data might be making your jobs fail if your application logic does not handle it well.

In this section, you have learned how you can debug or troubleshoot a slow-running cluster and in the next section, you will learn about logging in your EMR cluster that includes the different default log files available and how you can archive or aggregate them in Amazon S3.

Logging in your EMR cluster

In the previous section, we explained how you can leverage log files available in your cluster to debug a cluster failure or slow-running jobs. In this section, we will dive a bit more deeply into logging to explain what different log files are available in each path of the cluster and how you can integrate log archiving with Amazon S3.

Default log files available in EMR

By default, EMR clusters are configured to write log files to the `/mnt/var/log` directory of the master node, and to access them you can SSH to the master node. These log files are available till the time the master node is in the running state and if it terminates for any reason, then you lose access to these log files, so it's always a great idea to archive log files to Amazon S3 for persistence.

The following are the different types of log files generated by your cluster:

- **Step logs**: These logs include the result of each step and are stored in the `/mnt/var/log/hadoop/steps/` directory of the master node. Each step's log files are separated by a subdirectory that has a 13-character step identifier and includes an incremental number at the end to represent each step. For example, for step 1, the subdirectory path will be `/mnt/var/log/hadoop/steps/s-<stepId>1/` and `stepId` will be unique for the cluster.

- **Hadoop and YARN component logs**: These logs include different Hadoop and YARN component logs, which are available as subdirectories under `/mnt/var/log`. A few subdirectory examples include `hadoop-mapreduce`, `hadoop-yarn`, `hadoop-hdfs`, and `hadoop-httpfs`. There is an additional `hadoop-state-pusher` subdirectory, which stores the output of the Hadoop state pusher process.

- **Bootstrap action logs**: If you have configured bootstrap actions for your cluster, then its logs are stored under the `/mnt/var/log/bootstrap-actions/` directory of the master node. Each bootstrap action stores its log output in a separate subdirectory, which is an incremental number. For example, the first bootstrap action will have the path `/mnt/var/log/bootstrap-actions/1/`.

- **Instance state logs**: These logs provide EC2 instance-specific information that includes CPU, memory, and garbage collector threads of the cluster node and are stored in the `/mnt/var/log/instance-state/` directory of the master node.

As you have learned, all these logs are configured to store output in the cluster's master node by default. In the next section, let's understand how you can configure these logs to be archived to Amazon S3 for persistence.

Archiving log files to Amazon S3

While launching your cluster, you can define configuration to archive your master node logs to Amazon S3. By default, clusters launched using the EMR console have this setting enabled but clusters launched using the AWS CLI or the EMR API need to have it enabled. These logs are pushed to Amazon S3 every 5 minutes and there is a chance the last 5 minutes of log data will not be pushed to Amazon S3 when the cluster gets terminated.

Let's look at a few options to configure Amazon S3 log archival or aggregation.

- **Archive logs to Amazon S3 using the AWS Console**: When you launch a cluster with the EMR console, logging is enabled by default for both quick create and advanced options with a default Amazon S3 log path that you can change. But for the advanced options, you get an additional configuration to enable encryption for your log files, where you can specify the ARN of your AWS KMS key. This encryption option is available for clusters using EMR 5.30.0.

- **Archive logs to Amazon S3 using the AWS CLI**: To archive logs to Amazon S3, you can specify the `--log-uri` parameter while launching the cluster using the AWS CLI. The following is a sample command using the AWS CLI that specifies the Amazon S3 path through the `--log-uri` parameter:

```
aws emr create-cluster --name "Archive cluster log"
--release-label emr-6.3.0 --log-uri s3://<mybucket>/
logs/ --applications Name=Hadoop Name=Hive
Name=Spark --use-default-roles --ec2-attributes
KeyName=<myEC2KeyPair> --instance-type m5.2xlarge
--instance-count 3
```

This helps to archive logs into Amazon S3, but if you plan to aggregate a single application log to a single file, then you can look at the following configuration to aggregate logs.

Aggregating logs in Amazon S3 using the AWS CLI

When an application runs on the cluster, it gets executed as distributed tasks running in different cluster nodes where each node container generates its own log for that application. If you plan to aggregate these container logs to a single file, then while launching the cluster, you can specify additional configuration through bootstrap actions. This feature is available in EMR starting from EMR 4.3.0.

Let's assume we have saved this JSON configuration file as `s3-log-aggregation-config.json`:

```
[
  {
    "Classification": "yarn-site",
    "Properties": {
      "yarn.log-aggregation-enable": "true",
      "yarn.log-aggregation.retain-seconds": "-1",
      "yarn.nodemanager.remote-app-log-dir": "s3:\/\/<my-log-
bucket>\/logs"
    }
  }
]
```

Please replace <my-log-bucket> with your bucket name and then pass this configuration file while creating the cluster using the AWS CLI:

```
aws emr create-cluster --name "EMR Log Aggregation cluster"
--release-label emr-6.3.0 --applications Name=Hadoop
--use-default-roles --ec2-attributes KeyName=<myEC2KeyPairName>
--instance-type m5.xlarge --instance-count 3 --configurations
file://./s3-log-aggregation-config.json
```

Please replace <myEC2KeyPairName> with your EC2 key pair name.

Enabling the debugging tool

This additional debugging tool allows you to browse log files more easily from the EMR console. Enabling this option is available both from the EMR console's advanced cluster create option or through the AWS CLI. You need to enable logging to use this debugging tool.

When you enable debugging on your cluster, EMR archives the log files to S3 and indexes those files for easier access.

In the EMR console, when you choose the advanced cluster create option, you can find this option in the **General cluster settings** section's **General Options** configuration parameters. If you are using the AWS CLI to create the cluster, you should specify --enable-debugging with the --log-uri parameter.

Summary

Over the course of this chapter, we have got an overview of how you can create an EMR cluster using both the AWS console's quick and advanced creation options with different configuration options. We have also provided an overview of how you can integrate custom AMIs for your cluster and how termination protection can help for transient cluster use cases.

Finally, we covered the different logging and troubleshooting options you have to debug your cluster or job failures.

That concludes this chapter! Hopefully, you have got a good overview of setting up an EMR cluster with its different configurations and in the next chapter, we can dive deep into different monitoring, scaling, and high availability concepts.

Test your knowledge

Before moving on to the next chapter, test your knowledge with the following questions:

1. Assume on top of default EMR configurations, you need to install a few additional libraries and, post-installation, execute a few scripts. This process will be repeated every time a new instance is added to the cluster. How will you implement this while launching your cluster?

2. You have a running EMR cluster, where you have one Hive and one Spark job configured to be executed in a sequence as EMR steps. You have noticed that step 2, which is a Spark job, is failing. With further analysis, you have identified that all tasks of that Spark job are completed but one task is running for a long period of time, which makes the whole process slow. How will you resolve this problem?

3. Your organization has compliance policies that say all the application logs need to be persistent at least for a year. You are going to integrate EMR for one of your transient cluster use cases that will do batch ETL operations. To be compliant with your organization's policy, how should you configure your EMR cluster?

Further reading

Here are a few resources you can refer to for further reading:

*   Learn more about tagging a cluster: `https://docs.aws.amazon.com/emr/latest/ManagementGuide/emr-plan-tags.html`

*   Learn more about the networking of a cluster: `https://docs.aws.amazon.com/emr/latest/ManagementGuide/emr-plan-vpc-subnet.html`

*   Common errors in EMR: `https://docs.aws.amazon.com/emr/latest/ManagementGuide/emr-troubleshoot-errors.html`

6
Monitoring, Scaling, and High Availability

In the previous chapter, you learned how to set up your EMR cluster and configure it with advanced settings related to hardware, software, and security and how to troubleshoot failures or slow-running clusters. In this chapter, we will dive deeper into cluster **monitoring**, **scaling**, and **high-availability** features.

Scaling cluster resources is an important aspect as you don't need to manually resize the cluster and also size the cluster based on specific workloads. In this chapter, you will learn about the autoscaling and managed scaling capabilities of EMR and how Amazon CloudWatch monitoring plays a role in it.

The following are the high-level topics that we will cover in this chapter:

- Monitoring your EMR cluster
- Scaling cluster resources
- Comparing managed scaling with autoscaling
- Cluster cloning and high availability with multiple master nodes

Technical requirements

In this chapter, we will dive deep into EMR cluster monitoring, scaling, and high-availability aspects. To test out the features and configurations, you will need the following resources before you get started:

- An AWS account

- An IAM user that has permission to create an EMR cluster, EC2 instances, and dependent IAM roles and has access to CloudWatch, CloudTrail logs, and more

Now, let's dive deep into the EMR cluster's monitoring aspects, which includes web interfaces available for your cluster's big data applications and Amazon CloudWatch and CloudTrail logs.

Monitoring your EMR cluster

When you think about monitoring your Amazon EMR cluster, you can consider the following options:

- Using the EMR console to get the overall cluster status, the health of nodes, and the high-level status of YARN or Hadoop Spark applications

- Analyzing logs generated by EMR and your big data applications, which might be stored in the master node or core task nodes

- Accessing web interfaces of different Hadoop applications to analyze the job status or task execution or Ganglia to monitor the overall performance of your cluster

- Using Amazon CloudWatch for logging, monitoring, and integrating rule-based notifications

- Using Amazon CloudTrail to audit the access logs for your EMR cluster APIs

We covered the first two options in the previous chapter, where we explained how you can use the EMR console to monitor cluster status and how you can access logs available in the master node with the log archive to Amazon S3.

Now, let's dive deep into the remaining options that you can use for monitoring your cluster and jobs.

Monitoring clusters and applications with web user interfaces

As highlighted earlier, the EMR cluster provides access to a big data application's web interfaces, using which you can monitor the cluster. If you have configured **Ganglia** on your cluster, then you can use the Ganglia web interface to monitor your cluster's overall performance, the usage of memory, and the CPU. If you have configured **Spark** on your cluster, then you can use the Spark history server to monitor the execution of your jobs with the amount of resources or the time each task took to complete.

Before diving deep into these application user interfaces, let's understand what configuration steps you need to follow to access them.

Accessing a big data application's web interfaces hosted on EMR clusters

EMR clusters are configured with security measures so that all access goes through a defined authentication and authorization mechanism. When you configure Hadoop or other big data applications on your EMR cluster, for security reasons, it makes the respective application's web interface available on the cluster master node's local web server. So, if you wish to access them, then you need to connect to the master node using SSH and let the browser web request go through the authentication proxy to get access.

There are a few Hadoop applications that are also hosted as websites in core and task nodes and they are also available on the local web servers of those nodes.

EMR makes the web interfaces available through a specific port and you can access the web interfaces using the master node's public DNS or the core node's public DNS. The following list includes URLs for all the web interfaces that you can access in EMR:

- **Ganglia**: `http://master-public-dns-name/ganglia/`
- **Hadoop HDFS NameNode**: `http://master-public-dns-name:50070/`
- **Hadoop HDFS NameNode (EMR version pre-6.x)**: `https://master-public-dns-name:50470/`
- **Hadoop HDFS NameNode (EMR version 6.x)**: `https://master-public-dns-name:9871/`

- **Hadoop HDFS DataNode**: `http://coretask-public-dns-name:50075/`

- **Hadoop HDFS DataNode (EMR version pre-6.x)**: `https://coretask-public-dns-name:50475/`

- **Hadoop HDFS DataNode (EMR version 6.x)**: `https://coretask-public-dns-name:9865/`

- **HBase**: `http://master-public-dns-name:16010/`

- **Spark history server**: `http://master-public-dns-name:18080/`

- **Tez**: `http://master-public-dns-name:8080/tez-ui`

- **Flink history server (EMR version 5.33 and later)**: `http://master-public-dns-name:8082/`

- **Hue**: `http://master-public-dns-name:8888/`

- **JupyterHub**: `https://master-public-dns-name:9443/`

- **Zeppelin**: `http://master-public-dns-name:8890/`

- **Livy**: `http://master-public-dns-name:8998/`

- **YARN NodeManager**: `http://coretask-public-dns-name:8042/`

- **YARN ResourceManager**: `http://master-public-dns-name:8088/`

Please replace `<master-public-dns>` and `<coretask-public-dns>` with the relevant values in the preceding URIs. You can get your master node's public DNS from the EMR console's **Summary** tab.

The following are a few options using which you can access the preceding web interfaces:

- **Using the Lynx text-based browser**: You can SSH to your master node using the master node IP or public DNS, which you can get from the EMR console, and the EC2 key pair you configured while creating the cluster. **Lynx** is a text-based browser that cannot show graphics and you have limited options with it. As an example, you can use the following command to access the **Hadoop ResourceManager** URI from the Linux prompt of the master node but cannot view the graphical user interface:

  ```
  lynx http://ip-###-##-##-###.us-east-1.compute.
  internal:8088/
  ```

- **SSH tunneling with local port forwarding**: You can SSH to the master node and then configure SSH tunneling with local port forwarding, which will allow you to access the web interface using any web-based browser.

- **SSH tunneling with dynamic port forwarding**: This method is good for new users who are not very familiar with security configurations. Similar to other options, you can SSH to the master node and then configure SSH tunneling with dynamic port forwarding, which will require you to configure a browser-based plugin to access the web interfaces with specific ports. For example, for the Google Chrome browser, you can configure **SOCKS** proxy settings, and for the Firefox browser, you can configure the **FoxyProxy** add-on.

Out of the preceding three options, the last two SSH tunneling approaches are more popular as normal internet-based web browsers can support complete graphics-based web interfaces.

Before we dive deep into how you can do SSH tunneling, let's understand how you can configure your EMR cluster's security group to allow inbound SSH access, which is a prerequisite before doing SSH.

Before you connect – allowing inbound SSH traffic in an EMR cluster security group

Before connecting to the master node from your local system using SSH, you need to make sure that you have allowed inbound SSH access in your EMR cluster's security group. The following steps can guide you to configure this:

1. Navigate to the Amazon EMR console at `https://console.aws.amazon.com/elasticmapreduce/`.

2. From the **Clusters** list, select the name of the cluster that you plan to do SSH on.

3. In the **Summary** tab of the cluster, select the security groups for **Master** under the **Security and access** section.

4. Click on the **ElasticMapReduce-master** security group, which will take the EC2 service's security group list with the filter applied to your security group.

5. Select the **Inbound Rules** tab and then click **Edit inbound rules**.

6. Check for an inbound rule that has **Type** as **SSH**, **Port** as **22**, and **Source** as **0.0.0.0/0**. If the rule exists, choose **Delete** to remove it. Please note, before December 2020 the default **ElasticMapReduce-master** security group had a preconfigured rule that allowed inbound from all sources on port 22. The rule was added to simplify SSH access to the master node, but it is strongly recommended to remove it and add more restricted access that opens access to specific IPs or security groups.

To provide an overview of **Classless Inter-Domain Routing (CIDR)**, it improves the IP address allocation process by replacing the old A, B, C-based process. CIDR is a combination of two components: one is a typical IPv4 address and the second part is a suffix that represents how many bits there are in the entire address. CIDR `0.0.0.0/0` represents all the IP addresses, which means allowing permissions to all the source systems.

7. Then, scroll to the bottom of the rule list and select **Add Rule**.

8. Specify **Type as SSH**, which will auto populate **Protocol** as **TCP** and **Port** as **22**. For **Source**, select **My IP** from the list to restrict access to only your system. You can add additional rules if you would like to provide access to other IP ranges, but avoid choosing **Anywhere-IPv4** or **Anywhere-IPv6**, which will make it publicly accessible.

9. Then, click **Save**.

10. As an optional step, you can repeat the same process for the **ElasticMapReduce-slave** security group, if you would like to SSH to core or task nodes.

After allowing SSH access to your IP, we can see how to configure SSH tunneling to the EMR cluster's master node. To explain the detailed setup steps, I have taken an example of OpenSSH client software, which can guide you to look for options in other equivalent SSH client software, such as PuTTY for Windows users.

Setting up an SSH tunnel to the master node using local port forwarding

For this setup, you need to get your cluster master node's public DNS and EC2 key pair name that you will be using to do SSH tunneling to the master node. Then, to configure local port forwarding, you need to specify any unused local system port that will be used to forward the request traffic to a specific port of the target master node's local web server.

The following steps can guide you to configure the SSH tunneling:

1. Open a terminal window in OpenSSH. If you are using a Linux OS, it is available in **Applications | Accessories | Terminal**, and for macOS, it should be available under **Applications | Utilities | Terminal**.

2. As a next step, you need to execute the following ssh command, which uses the master node's public DNS and EC2 key pair for connection and also defines the target port you would like to connect to:

```
ssh -i ~/<EC2KeyPair>.pem -N -L 8999:ec2-###-##-##-###.
compute-1.amazonaws.com:8088 hadoop@ec2-###-##-##-###.
compute-1.amazonaws.com
```

3. In the preceding command, we have used 8999 as the local system's unused port, which you can replace, and it will forward this local port to the master node's 8088 port, which is used to access the **ResourceManager** web interface. Please replace the <EC2KeyPair> and ec2 hostnames before executing the command.

   The terminal remains open after you issue this command, and no answer is received.

4. Now, to access the **ResourceManager** web interface in your system browser, type http://localhost:8999/ in your browser address bar, which will forward the request to port 8088 of the master node.

5. To close the session, you can close the terminal window.

You can repeat these steps to access any other web interface available in the master node or any other core and task nodes. For example, to access the **JupyterHub** web interface, you can configure any other local unused port to forward the request to port 9443 of the master node.

Next, you will learn how to configure SSH tunneling with dynamic port forwarding.

Setting up an SSH tunnel to the master node using dynamic port forwarding

This method is similar to SSH tunneling with local forwarding with additional configuration to make the port forwarding dynamic so that you can get the benefit of the FoxyProxy or SwitchyOmega add-ons to manage your dynamic SOCKS proxy configurations.

SOCKS proxy management tools provide features to configure the automatic filtering of URLs based on text patterns, where you can specify the master node's URL pattern for matching. These SOCKS proxy-based browser plugins can automatically turn on or off when you switch between the master node URL and other website URLs.

After you have allowed SSH inbound access to your cluster security group and collected the master node DNS and EC2 key pair, you can take the following steps for the setup:

1. Open a terminal window in OpenSSH.

2. Next, execute the following `ssh` command, which uses the master node's public DNS and EC2 key pair for connection and also defines the local unused port, which will be used to forward requests to all the ports of the target master node's local web server:

```
ssh -i ~/<EC2KeyPairName>.pem -N -D 8999 hadoop@ec2-###-
##-##-###.compute-1.amazonaws.com
```

3. As you can see from the preceding command, it uses the `-D` option, which represents dynamic port forwarding, which enables a local SOCKS proxy listening on the local unused port you have specified in the command.

4. Next, you can configure the SOCKS proxy on your browser to access the web interfaces using the master or core node's public DNS and the respective web interface's port number. For example, you can access **Spark HistoryServer** using `http://master-public-dns-name:18080/`.

5. When you are done accessing the server, you can close the terminal window.

The preceding steps summarize how you can configure SSH tunneling to the master node using dynamic port forwarding. Next, we will understand how to configure the SOCKS proxy plugin on your browser.

Configuring the SOCKS proxy on your browser

While configuring the SOCKS proxy as a plugin or add-on, make sure to include the following configurations:

- Use localhost as the host address and the same local unused port you specified while setting up port forwarding in the terminal window.

- Specify SOCKS v5 as the protocol that will optionally allow you to set user authorization.

- Specify the following URL wildcard patterns as allowed:

  - To match a US Region-specific public DNS, use the `*ec2*.amazonaws.com*` and `*10*.amazonaws.com*` patterns.

  - To match all other Regions' public DNS, use the `*ec2*.compute*` and `*10*.compute*` patterns.

  - To access JobTracker log files, use the `10.*` pattern.

- To match the private or internal DNS names of a cluster in the us-east-1 Region, use the *.ec2.internal* pattern, and for all other Regions, use the *.compute.internal* pattern.

Based on the browser and add-on you are using, the steps for configuring the preceding URLs might vary. The following is an example if you need to configure it with the Google Chrome browser and SwitchOmega add-on:

1. Navigate to Google Chrome's extensions page, which is typically accessible through https://chrome.google.com/webstore/category/extensions, and then search for Proxy SwitchyOmega. Once found, click **Add to Chrome**.

2. On the plugin page, select **New profile** and specify emr-socks-proxy as the profile name.

3. Select **PAC profile** and then click **Create**. The **Proxy Auto-Configuration (PAC)** files allow you to configure a list of browser requests as an **allow list**, which should be forwarded to a proxy server.

4. Within the **PAC Script** field, replace the contents with the following, which uses port 8999 for forwarding requests to your proxy server. Please replace 8999 with the local unused port you have configured while setting up the SSL tunnel:

```
function FindProxyForURL(url, host) {
    if (shExpMatch(url, "*ec2*.amazonaws.com*")) return
'SOCKS5 localhost:8999';
    if (shExpMatch(url, "*ec2*.compute*")) return 'SOCKS5
localhost:8999';
    if (shExpMatch(url, ""http://10.*")) return 'SOCKS5
localhost:8999';
    if (shExpMatch(url, "*10*.compute*")) return 'SOCKS5
localhost:8999';
    if (shExpMatch(url, "*10*.amazonaws.com*")) return
'SOCKS5 localhost:8999';
    if (shExpMatch(url, "*.compute.internal*")) return
'SOCKS5 localhost:8999';
    if (shExpMatch(url, "*ec2.internal*")) return 'SOCKS5
localhost:8999';
    return 'DIRECT';
}
```

5. Under the **Actions** left-panel navigation, select **Apply changes** to save your proxy settings.

6. As a final step, on your Google Chrome toolbar, select the **SwitchyOmega** plugin and the **emr-socks-proxy** profile that you configured.

7. Validate your configuration by accessing `http://master-public-dns-name:18080/`.

In this section, we saw in detail how to configure an SSH tunnel to the master node and how to access web interfaces of Hadoop applications available in your EMR cluster. Next, we will look at a couple of Hadoop interfaces that you can use to monitor your cluster.

Viewing cluster performance metrics with Ganglia

Ganglia is an open source project that is scalable and designed to monitor usage and performance metrics of distributed clusters or grids. You can set up and integrate Ganglia on your cluster to monitor the performance of individual nodes and the cluster as a whole. Ganglia is available in EMR starting from the 4.2 release.

In an EMR cluster, Ganglia is configured to capture and visualize Hadoop and Spark metrics. It provides a web interface where you can see your cluster performance with different graphs and charts representing CPU and memory utilization, network traffic, and loading of the cluster. As explained in the previous section, you can access the web interface through `http://master-public-dns-name/ganglia/` by configuring SSH tunneling to the master node.

Monitoring cluster metrics with CloudWatch monitoring

Amazon EMR publishes different cluster- and job-level metrics to **Amazon CloudWatch**, which can be used to configure rules for event-based notifications and dashboards for monitoring. For example, you can configure rules to react when a cluster status changes from **WAITING** to **RUNNING** or your cluster master node's CPU usage goes beyond a certain threshold.

In the following sub-section, you will learn how to monitor CloudWatch events and metrics.

Monitoring CloudWatch events

Amazon EMR automatically tracks and publishes events as JSON objects to an event stream of CloudWatch. These events include changes in cluster states, instance groups, autoscaling policies, and changes in steps. Each event includes information such as details about the event, the date and time of the event, the EMR cluster, or the instance group affected by the event.

Every time EMR publishes an event, it also includes the severity of the event and the event message. For example, the following table lists cluster events with the state, severity, and message:

State or State Change	Severity	Message
STARTING	INFO	Amazon EMR cluster ClusterId (ClusterName) was requested at Time and is being created.
RUNNING	INFO	Amazon EMR cluster ClusterId (ClusterName) began running steps at Time.
WAITING	INFO	Amazon EMR cluster ClusterId (ClusterName) was created at Time and is ready for use. —OR— Amazon EMR cluster ClusterId (ClusterName) finished running all pending steps at Time.
TERMINATED	CRITICAL if the cluster is terminated with INTERNAL_ERROR, VALIDATION_ERROR, INSTANCE_ FAILURE, BOOTSTRAP_FAILURE, or STEP_FAILURE state change reasons INFO if the cluster is terminated with either ALL_STEPS_COMPLETED or USER_REQUEST state change	Amazon EMR cluster ClusterId (ClusterName) has terminated with errors at Time with a reason of StateChangeReason:Code.
TERMINATED_ WITH_ERRORS	CRITICAL	Amazon EMR cluster ClusterId (ClusterName) has terminated with errors at Time with a reason of StateChangeReason:Code.

Table 6.1 – A table showing event state, severity, and message for an EMR cluster

Similar to each step change event, instance fleet or instance group events and autoscaling events also get pushed to CloudWatch, which you can use for monitoring.

Now let's learn how you can view these events in your EMR console.

Viewing events using the EMR console

For each cluster, you can see a simple list of events being published on the EMR console in descending order.

You have an option to view events of all the clusters available in a Region in descending order. You can navigate to the EMR console and select the **Events** sub-menu navigation to see the events of all clusters.

If you want to restrict access where you don't want a user to view all cluster events for any Region, then you can add the `"Effect": "Deny"` statement for the `elasticmapreduce:ViewEventsFromAllClustersInConsole` action to a policy and then attach it to the IAM user.

The following steps explain how you can view events for a single cluster:

1. Navigate to the EMR console.

2. Select **Clusters** under **EMR on EC2**.

3. From the cluster list, select the cluster for which you want to see the events.

4. Then, select the **Events** tab, which will list events, as shown in the following screenshot:

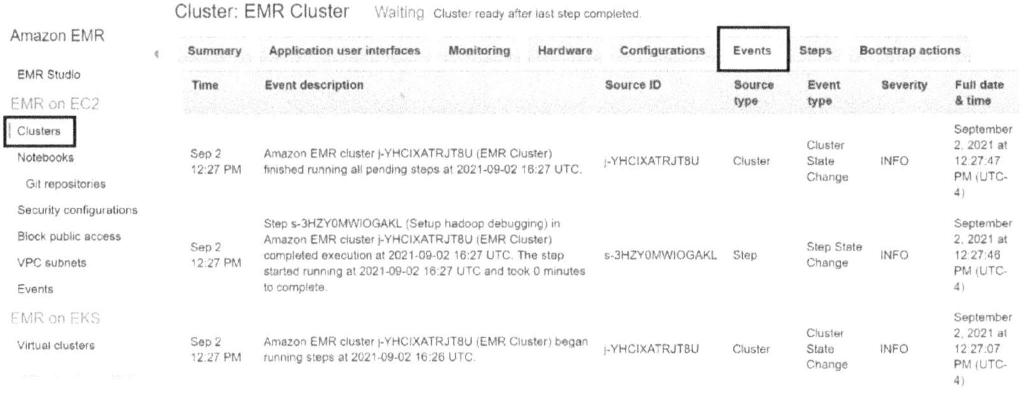

Figure 6.1 – List of events of a specific cluster in the EMR console

Now, let's learn how to create CloudWatch event rules.

Creating CloudWatch event rules for EMR events

As described earlier, EMR sends each event as an event stream JSON object to CloudWatch and you can configure rules in CloudWatch for any of the JSON attributes. The CloudWatch rule might have a pattern matching rule defined for the source event and will have the target configured. CloudWatch supports integration with several AWS services, so you can configure the target action as sending an SNS notification or trigger an AWS Glue or AWS Lambda job.

The following is a sample event of an EMR cluster when the status changes to TERMINATED:

```
{
    "version": "0",
```

```
   "id": "1234abb0-f87e-1234-b7b6-000000123456",
   "detail-type": "EMR Cluster State Change",
   "source": "aws.emr",
   "account": "<AWS-account>",
   "time": "2021-12-16T21:00:23Z",
   "region": "us-east-1",
   "resources": [],
   "detail": {
      "severity": "INFO",
      "stateChangeReason": "{\"code\":\"USER_
REQUEST\",\"message\":\"Terminated by user request\"}",
      "name": "Dev Cluster",
      "clusterId": "j-123456789ABCD",
      "state": "TERMINATED",
      "message": "Amazon EMR Cluster jj-123456789ABCD (Dev
Cluster) has terminated at 2021-12-16 21:00 UTC with a reason
of USER_REQUEST."
   }
}
```

Next, let's learn about monitoring metrics.

Monitoring CloudWatch metrics

You can view the metrics Amazon EMR publishes to CloudWatch using the CloudWatch or Amazon EMR console. If you have your own custom logging and monitoring solutions or are looking to build automation around these metrics, then you can retrieve these metrics' data using the `mon-get-stats` CloudWatch CLI command or the `GetMetricStatistics` CloudWatch API.

The following steps will explain how you can view the metrics in your EMR console:

1. Navigate to the Amazon EMR console at `https://console.aws.amazon.com/elasticmapreduce/`.

2. From the **Clusters** list, select the cluster for which you plan to view the metrics.

3. Select the **Monitoring** tab, which will have subtabs of **Cluster Status**, **Node Status**, and **IO**. Select any of the tabs to view metrics reports for your cluster or nodes.

4. After selecting any of the subtabs, you can select the graph size and start and end fields to filter the metrics data by a specific time frame.

5. Once you have metrics data available in CloudWatch, you can define alarms based on specific parameter thresholds. For example, if HDFS utilization goes beyond 90%, then send an email to your cloud administrator team.

Let's take a look at some of the metrics that are reported by Amazon EMR.

Metrics reported by EMR to CloudWatch

Amazon EMR automatically sends metrics data to CloudWatch every 5 minutes, which is archived for 2 weeks and then discarded. EMR pulls metrics from the cluster, so if the cluster is not reachable, no metrics are reported till the cluster is available.

EMR publishes several metrics to CloudWatch. The following list provides an example of a few metrics:

- **Amazon EMR metrics (AWS/ElasticMapReduce namespace)**: If your cluster runs Hadoop 2.x, then a few of the example metrics it publishes are `IsIdle`, `ContainerAllocated`, `ContainerReserved`, `ContainerPending`, `AppsRunning`, and `CoreNodesRunning`.

- **Cluster capacity metrics**: When you have managed scaling enabled, EMR publishes various metrics that represent your cluster's current and target capacity. `TotalUnitsRequested`, `TotalUnitsRunning`, `CoreUnitsRequested`, and `CoreUnitsRunning` are a few examples of the metrics reported for it.

You can filter Amazon EMR metrics by either `JobFlowId` or `JobId`. Now, `JobFlowId` is the same as your cluster ID, which has a format of `j-XXXXXXXXXXXXX`, and `JobId` represents a specific job's ID that is in the format `job_XXXXXXXXXXXX_XXXX`. After learning how you can use Amazon CloudWatch for monitoring your cluster-, node-, and job-level metrics, let's look at Amazon CloudTrail, which can help in auditing user actions in AWS.

EMR API audit logging with AWS CloudTrail

AWS CloudTrail is a popular service that enables the continuous logging of AWS API activities and can help in operational or risk auditing, maintaining governance, and compliance. Every action you take on your EMR cluster, using the AWS console, AWS CLI, or EMR API, is logged to Amazon CloudTrail as an activity event. You can create a trail in CloudTrail, which will allow you to enable the continuous delivery of CloudTrail log events to Amazon S3. Even if you don't configure a trail, you can still see the most recent activity in **Event History** of the CloudTrail console.

AWS CloudTrail is enabled by default when you create an AWS account and its activity logs facilitate implementing security alarms by detecting unusual activity in your AWS account. Amazon EMR also integrates with AWS CloudTrail where each EMR console activity, AWS CLI command, or API invocation is logged to CloudTrail.

EMR information in AWS CloudTrail

When an activity occurs in Amazon EMR, it gets recorded in CloudTrail **Event History**. You have the option to view, download, or search recent events in your AWS account. It is always recommended to create a trail in CloudTrail, because of the following benefits:

- You can enable ongoing delivery of CloudTrail events as log files to Amazon S3. This helps you plan to persist your logs beyond 90 days.

- Optionally, you can configure alarms by pushing log events to Amazon CloudWatch.

- Also, you can query the event log data with SQL using Amazon Athena's query engine.

Every log entry or event contains the following information:

- What is the request (AWS Region, AWS service, and its API)?

- Who made the request?

- What is the source IP?

- When was it made?

- Other additional entries

To identify who made the request, CloudTrail provides the following information as part of the `userIdentity` property:

- Whether the request was made with an AWS IAM user

- Whether the request was made by a federated user with temporary security credentials

- Whether it's another AWS service that submitted the request; you can identify that from the `principalId` key in the JSON file

The following is an example JSON CloudTrail event that was logged for the terminate cluster action from the EMR console:

```
{
    "eventVersion": "1.08",
    "userIdentity": {
        "type": "AssumedRole",
        "principalId": "<ID>",
        "arn": "<ARN>",
        "accountId": "<AWS-Account-ID>",
        "accessKeyId": "<Access-Key>",
        "sessionContext": {
            "sessionIssuer": {
                "type": "Role",
                "principalId": "<Principal-ID>",
                "arn": "<ARN>",
                "accountId": "<AWS-Account-ID>",
                "userName": "developer"
            },
            "webIdFederationData": {},
            "attributes": {
                "creationDate": "2021-09-02T21:30:40Z",
                "mfaAuthenticated": "false"
            }
        }
    },
    "eventTime": "2021-09-02T21:32:31Z",
    "eventSource": "elasticmapreduce.amazonaws.com",
    "eventName": "TerminateJobFlows",
    "awsRegion": "us-east-1",
    "sourceIPAddress": "52.95.4.21",
    "userAgent": "AWS ElasticMapReduce Console",
    "requestParameters": {
        "jobFlowIds": [
            "j-<id>"
        ]
    },
```

```
    "responseElements": null,
    "requestID": "1e9e53be-31ec-4ed2-9c04-68714dac7e6a",
    "eventID": "403b5ef1-bf9b-436e-8d34-58e4d632fbf9",
    "readOnly": false,
    "eventType": "AwsApiCall",
    "managementEvent": true,
    "recipientAccountId": "<AWS-Account-ID>",
    "eventCategory": "Management"
}
```

In this section, we covered monitoring aspects of your cluster, jobs using Hadoop web interfaces, Amazon CloudWatch, and AWS CloudTrail. Next, we will look at the options EMR provides to scale cluster resources.

Scaling cluster resources

When you launch an Amazon EMR cluster for big data processing, most of the time, the computing capacity you need for your jobs is different. The number of resources you need for your cluster depends on the data volume of the file size, the kind of processing logic you have, and whether your cluster resources are being shared by any other jobs.

There are a few cases where you have defined a data volume and you are able to do capacity planning to launch a fixed node cluster that does not need any scaling capacity. But in most cases, you will have a variable workload or a shared cluster for multiple workloads that needs to react to on-demand capacity needs, where you will need to scale your cluster capacity dynamically.

Amazon EMR provides flexibility to configure the scaling of cluster resources as it provides two scaling features, that is, **EMR-managed scaling** and **autoscaling with a custom scaling policy**. When considering automatic scaling of your cluster, please take note of the following considerations:

- Your EMR cluster can have one or three master nodes, which you configure when you launch your cluster. You cannot apply scaling to master nodes and change the number of master nodes after the cluster is launched.

- You can change the configuration of your instance group after the cluster is launched, but you cannot apply scaling or resize your instance group when any reconfiguration is initiated. Also, you cannot change configurations when any resize is triggered through scaling.

Now, let's dive deep into both the scaling features and understand how they are different from each other.

Managed scaling in EMR

You can enable EMR-managed scaling starting from the EMR 5.30.0 release, except for EMR 6.0.0. EMR-managed scaling automates the cluster resource scaling without expecting you to configure any scaling rules. It evaluates cluster metrics continuously to make scaling decisions, which will help you to optimize cluster resource usage based on the need and can provide you with cost savings or better performance.

While enabling managed scaling, you need to set the following parameters, which provides a minimum and maximum range for the core and task node instances:

- **Minimum** (`MinimumCapacityUnits`): This is the minimum EC2 capacity that needs to be maintained in the cluster. It is measured through vCPUs; for instance fleets, it is measured through units, and for instance groups, it is measured through instances.

- **Maximum** (`MaximumCapacityUnits`): This is the maximum EC2 capacity that needs to be maintained in the cluster. Similar to minimum capacity units, it is measured through vCPUs; for instance fleets, it is measured through units, and for instance groups, it is measured through instances.

- **On-demand limit** (`MaximumOnDemandCapacityUnits` – optional): This is an optional parameter that specifies the maximum on-demand type EC2 capacity that can be added to the cluster. If not specified, it takes the default value of `MaximumCapacityUnits`.

  This parameter helps in deciding how many on-demand and spot instances will be included in the cluster capacity. For example, if you specify the minimum parameter as 10 instances, the maximum parameter as 50 instances, and on-demand maximum instances as 20, then EMR-managed scaling will scale on-demand instances up to 20, and the remaining 30 will be fulfilled with spot instances.

- **Maximum core nodes** (`MaximumCoreCapacityUnits` – optional): This is another optional parameter that represents the maximum allowed core node type capacity in the cluster. If not specified, it takes the default value of `MaximumCapacityUnits`.

> **Important Note**
>
> EMR-managed scaling is integrated to only work with YARN applications such as Hadoop, Spark, Flink, and Hive. At the time of writing this book, it does not support non-YARN-based applications, such as Presto.

Configuring managed scaling for your EMR cluster

You can configure managed scaling on your EMR cluster using the Amazon EMR console, AWS CLI commands, or AWS SDKs. Let's understand how you can enable managed scaling for your cluster using the console and CLI commands.

Enabling managed scaling using the AWS console

When you launch your Amazon EMR cluster using the EMR console, you can enable managed scaling with both quick create and advanced options.

With the quick create option, under **Hardware configuration**, you have the **Cluster scaling** option, which you can check to enable. It allows you to configure minimum (`MinimumCapacityUnits`) and maximum (`MaximumCapacityUnits`) core task nodes.

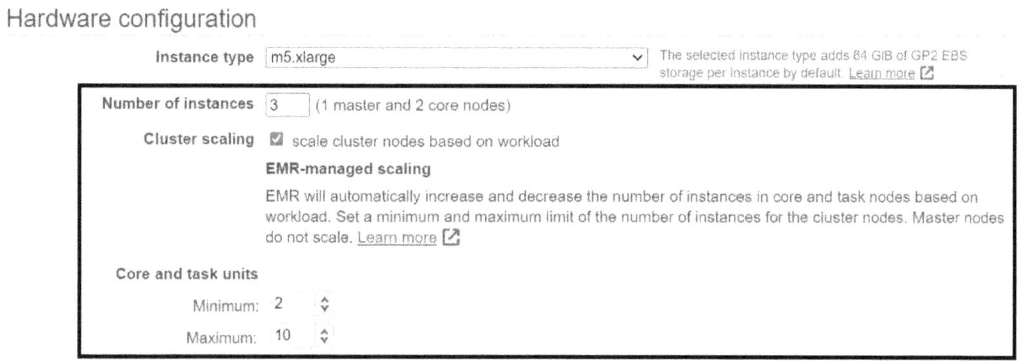

Hardware configuration

Instance type	m5.xlarge ⌄ The selected instance type adds 64 GiB of GP2 EBS storage per instance by default. Learn more
Number of instances	3 (1 master and 2 core nodes)
Cluster scaling	☑ scale cluster nodes based on workload
	EMR-managed scaling
	EMR will automatically increase and decrease the number of instances in core and task nodes based on workload. Set a minimum and maximum limit of the number of instances for the cluster nodes. Master nodes do not scale. Learn more
Core and task units	
Minimum:	2
Maximum:	10

Figure 6.2 – EMR console's quick cluster creation screen that shows cluster scaling configuration

As you can see in the preceding screenshot, the EMR cluster's quick create option does not include on-demand limit and maximum core node options, which are only available for the advanced cluster create option.

When you create a cluster using advanced options, on the **Step 2: Hardware** page, **Cluster scaling** is a separate section that allows you to configure EMR-managed scaling or a custom autoscaling policy. The following is a screenshot of the EMR console that represents the scaling options:

Cluster scaling

Adjust the number of Amazon EC2 instances available to an EMR cluster via EMR-managed scaling or a custom automatic scaling policy. Learn more ⧉

	Cluster scaling	✔	Enable Cluster Scaling
		●	Use EMR-managed scaling
			Create a custom automatic scaling policy

EMR-managed scaling

EMR will automatically increase and decrease the number of instances in core and task nodes based on workload. Set a minimum and maximum limit of the number of instances for the cluster nodes. Master nodes do not scale.

Core and task units

Minimum:	2	↕
Maximum:	4	↕
On-demand limit :	4	↕
Maximum Core Node :	4	↕

Figure 6.3 – EMR console's advanced cluster creation screen that shows managed scaling configuration

As you can see in the preceding screenshot, you have the **On-demand limit** (`MaximumOnDemandCapacityUnits`) and **Maximum Core Node** (`MaximumCoreCapacityUnits`) configuration parameters available, which were missing in the quick create option.

Changing scaling configuration for an existing running cluster

You can modify the scaling configurations for an already running cluster. The following steps will guide you on how to do it:

1. Navigate to the Amazon EMR console at `https://console.aws.amazon.com/elasticmapreduce/`.

2. From the **Clusters** list, select the cluster for which you plan to change the configuration.

3. Select the **Hardware** tab.

4. Click **Edit** under the **Cluster scaling** section.

5. Modify the maximum or minimum values as needed.

In this section, you learned how to configure managed scaling while creating a cluster or change the configuration for an already running cluster using the AWS console. In the next section, you will learn how to enable managed scaling using the AWS CLI.

Enabling managed scaling using the AWS CLI

When you launch your Amazon EMR cluster using the EMR console, you can enable managed scaling with both **Quick create** and **Advanced** options. Let's take a look at the command:

```
aws emr create-cluster \
service-role EMR_DefaultRole \
  --release-label emr-6.3.0 \
  --name EMR_Managed_Scaling_Cluster \
  --applications Name=Spark Name=Hbase \
  --ec2-attributes KeyName=<EC2KeyPairName>,InstanceProfile=EMR_
EC2_DefaultRole \
  --instance-groups InstanceType=m4.
xlarge,InstanceGroupType=MASTER,InstanceCount=1
InstanceType=m4.xlarge,InstanceGroupType=CORE,InstanceCount=2 \
  --region us-east-1 \
  --managed-scaling-policy
ComputeLimits='{MinimumCapacityUnits=2,
MaximumCapacityUnits=4,UnitType=Instances}''
```

Please replace `<EC2KeyPairName>` with the appropriate value before executing the preceding command. As you can see, there is an additional `managed-scaling-policy` added to the end of the command that specifies the configurations for scaling.

You can also enable managed scaling for an already running cluster using the AWS CLI. The following shows an example of this, where the `emr put-managed-scaling-policy` command enables the configuration:

```
aws emr put-managed-scaling-policy
--cluster-id <ClusterID>
--managed-scaling-policy
ComputeLimits='{MinimumCapacityUnits=1,
MaximumCapacityUnits=10,  MaximumOnDemandCapacityUnits=10,
UnitType=Instances}'
```

Alternatively, if you would like to disable managed scaling from a cluster, you can refer to the following command:

```
aws emr remove-managed-scaling-policy --cluster-id <ClusterID>
```

Please replace `<ClusterID>` with your existing cluster ID.

Understanding the node allocation strategy of managed scaling

EMR-managed scaling follows a scale-up and scale-down strategy to scale cluster resources automatically. In the following subsections, we will see what node allocation strategy EMR follows while scaling up or down.

Scale-up node allocation strategy

When you have configured all four parameters (maximum, minimum, maximum core nodes, and on-demand limit), EMR-managed scaling follows an order of assigning instances when it scales up. The following points explain how EMR-managed scaling allocates new instances:

- It first adds instances or capacity to core nodes as they have HDFS and Hadoop-related services set up, and then it adds the rest of the capacity to task nodes till the desired capacity is met.

- If you have set the `MaximumCoreCapacityUnits` parameter, then EMR scaling adds capacity to core nodes up to the maximum capacity allowed for core nodes, and then the rest is added to task nodes.

- If you have set the `MaximumOnDemandCapacityUnits` parameter, then while adding nodes, EMR scaling adds on-demand EC2 instance types till the maximum number is met, and then the rest of the capacity is fulfilled with spot instances.

- If both the `MaximumCoreCapacityUnits` and `MaximumOnDemandCapacity Units` parameters are set, then it considers both for node allocation.

  As an example, if `MaximumCoreCapacityUnits` is less than `MaximumOnDemandCapacityUnits`, then EMR will first scale core nodes up to the maximum core capacity limit. Then, for the rest of the capacity, it will use on-demand EC2 instances to scale the task nodes, up to the maximum limit defined for on-demand instances, and then the remaining task nodes are added using spot instances.

After understanding how EMR-managed scaling allocates nodes for scaling up, let's understand how it works for scaling-down scenarios.

Scale-down strategy

When your submitted jobs are completed, your cluster does not need to maintain the higher capacity and can work to scale down the resources so that your cost is low. EMR-managed scaling takes the following strategy to scale down:

- Opposite to the scaling-up approach, while scaling down, EMR first removes task nodes and then removes core nodes up to the minimum core node capacity limit of your AWS account, as it needs to maintain the minimum capacity.

- For both core and task node types, EMR first removes spot instances and then the on-demand instances.

While scaling down, if EMR receives a heavy load that needs resources, then it cancels the scale-down operation and adds capacity to scale up.

Next, we will learn about different CloudWatch metrics that help in making managed scaling decisions.

Understanding managed scaling CloudWatch metrics

When you have managed scaling enabled on your cluster, Amazon EMR publishes high-resolution metrics every 1 minute. You can view these metrics in both the EMR console and the Amazon CloudWatch console, which shows events for every scale-up or scale-down resize operation.

There are several metrics published by Amazon EMR and the following shows a few examples of them:

- **Current or target capacity-related metrics**: EMR publishes metrics that represent the current capacity and the target capacity. A few of the metrics are `TotalNodesRequested`, `TotalNodesRunning`, `CoreNodesRequested`, `CoreNodesRunning`, `TaskNodesRequested`, and `TaskNodesRunning`.

- **Usage status of cluster and jobs**: EMR publishes metrics at every 1-minute granularity, so you can relate these cluster and job status-related metrics with cluster capacity-related metrics to understand how scaling decisions are made with EMR-managed scaling. A few of the metrics published are `AppsPending`, `AppsRunning`, `ContainerAllocated`, `ContainerPending`, `ContainerPendingRatio`, `MemoryAvailableMB`, `YARNMemoryAvailablePercentage`, `HDFSUtilizatioin`, and `IsIdle`.

Using the CloudWatch console, you can create a graph that shows how the scale-up and scale-down operations are happening on your cluster. To create a graph, you can refer to the following steps:

1. Navigate to the Amazon CloudWatch console and click **Metrics** under **All metrics**.

2. From the **AWS namespaces** service list, select **EMR**, and then click **Job Flow Metrics**.

Then, apply a filter on **JobFlowId**, which is the EMR cluster ID, and **Metric Name**. You can select the checkbox next to each metric to show one or more metrics on the graph. The following is a screenshot of CloudWatch that shows how the number of nodes and container pending parameters changed when EMR-managed scaling requested additional nodes to handle a higher volume of data processing:

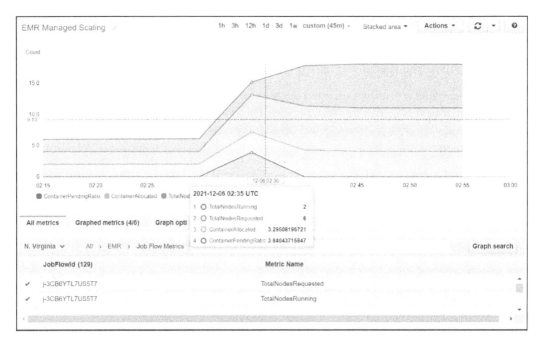

Figure 6.4 – CloudWatch metrics graph, showing EMR nodes scaling up by managed scaling

After understanding how EMR-managed scaling works, let's understand how you can define custom autoscaling policies if needed.

Autoscaling in EMR with a custom policy for instance groups

You can integrate cluster autoscaling with a custom policy starting with the EMR 4.0 release. This was the first scaling mechanism available in EMR, where you can define your own scale-up, scale-down criteria. EMR-managed scaling was introduced later, where intelligence was built to handle the scaling automatically. But the custom scaling policy option is still available, which provides more flexibility if you want to define your own rules for scaling up or down. You can configure scaling rules based on CloudWatch and other metrics published by EMR.

> **Important Note**
> Please note, scaling with a custom policy is only available to instance groups of EMR and is not available for instance fleets.

You can specify a custom scaling policy on an instance group while you are creating the cluster or after the cluster is launched. Each instance group (except the master) can have its own custom scale-up, scale-down rules.

You can define the rules using the AWS console, AWS CLI, and EMR API. While using the AWS CLI or EMR API, you need to provide the policy configuration using a JSON file. Also, while using the AWS CLI or EMR API, you can define a policy using a custom CloudWatch metric, which is not available in the AWS console.

When you initially launch a cluster with custom scaling policies, a default policy is preconfigured to get started that is suitable for many applications. But you do have the flexibility to modify or delete the default rules.

Prerequisites for configuring autoscaling with a custom policy

Before you configure a custom scaling policy on your cluster, please consider the following prerequisite steps:

- When you launch your EMR cluster, you must set the `VisibleToAllUsers` parameter to `true` for the autoscaling to work.

- When you create an EMR cluster using the AWS console, it automatically creates the `EMR_AutoScaling_DefaultRole` IAM role with the `AmazonElasticMapReduceforAutoScalingRole` policy attached to it. This role provides permission to add or terminate instances during a scaling operation. But if you are using the AWS CLI or EMR API, then you create either the default `EMR_AutoScaling_DefaultRole` by executing the `create-default-role` command or your custom role. After that, you can specify the role name to your –`auto-scaling-role` parameter.

After you have met the prerequisites and launched the cluster with autoscaling, we can learn how you can define configurations for scaling.

Configuring autoscaling rules

You can configure a custom autoscaling policy using both the EMR console and AWS CLI commands. When a scaling operation is triggered and the EC2 instance gets added to the cluster, they can be used by Hadoop services such as Hive, Spark, and Presto as soon as the instances reach the `InService` state.

For scaling in, between EMR releases 5.1.0 and 5.9.1, you have the option to specify how the instance gets terminated: either it is at task completion or it's at the EC2 instance-hour boundary for billing. But after EMR release 5.10.0, you don't have this option available and the default scale-in behavior is at task completion.

You cannot apply scaling rules to master nodes, so let's learn how you can specify the scale-out and scale-in rules for core and task nodes.

Configuring a custom scaling policy using the EMR console

When you launch your cluster using the EMR console's advanced options, under the **Hardware configuration** page, you have the **Cluster scaling** section, which can help you configure autoscaling with custom policies. The following screenshot shows how you can enable autoscaling with custom policies for the advanced cluster creation option:

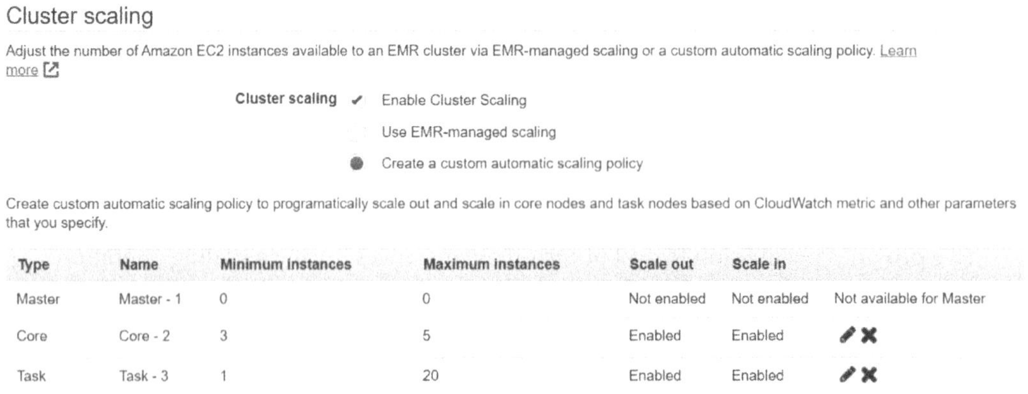

Cluster scaling

Adjust the number of Amazon EC2 instances available to an EMR cluster via EMR-managed scaling or a custom automatic scaling policy. Learn more [↗]

Cluster scaling ✓ Enable Cluster Scaling

Use EMR-managed scaling

⬤ Create a custom automatic scaling policy

Create custom automatic scaling policy to programatically scale out and scale in core nodes and task nodes based on CloudWatch metric and other parameters that you specify.

Type	Name	Minimum instances	Maximum instances	Scale out	Scale in	
Master	Master - 1	0	0	Not enabled	Not enabled	Not available for Master
Core	Core - 2	3	5	Enabled	Enabled	✎ ✖
Task	Task - 3	1	20	Enabled	Enabled	✎ ✖

Figure 6.5 – EMR console's advanced cluster creation screen that shows a custom automatic scaling policy

As you can see from the preceding screenshot, you can only configure rules for core and task nodes, where you can specify the minimum and maximum nodes for that specific instance type. The following screenshot shows how you can configure the scale-out and scale-in rule for them:

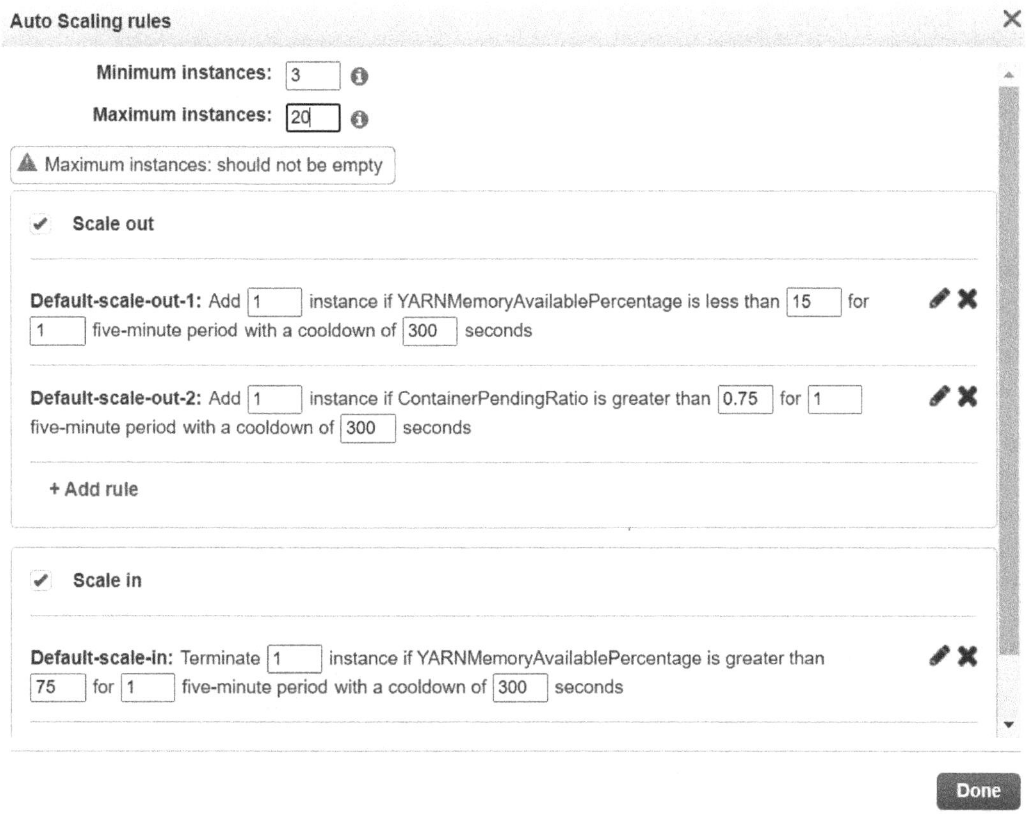

Figure 6.6 – EMR console screen that allows adding scale-out, scale-in rules

As you can see from the screenshot, you can configure multiple scale-out and scale-in rules to control the scaling behavior. Each rule is configured with the following parameters:

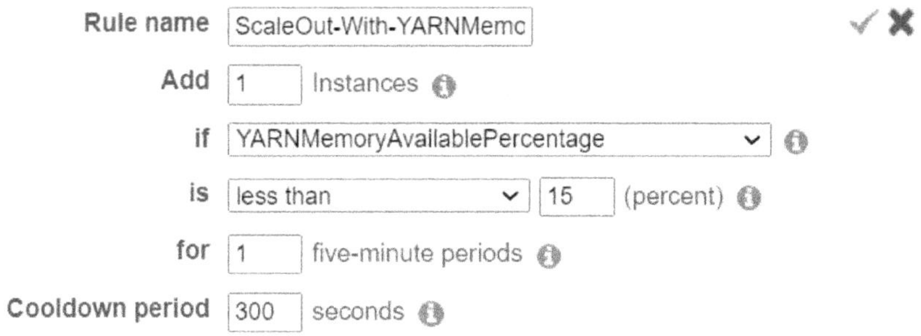

Figure 6.7 – EMR autoscaling rule parameters

> **Important Note**
> It is highly recommended to configure both scale-up and scale-down policies when you configure autoscaling with a custom policy, which will optimize your cluster's resource utilization, which can provide better performance and cost savings. Defining either scale out or scale in without the other one will require you to take manual action.

Configuring a custom scaling policy using the AWS CLI

When you launch a cluster using an AWS CLI command that needs to have autoscaling configured, you have the option to pass the JSON configuration directly in the command or save it in a JSON file and specify the file path in the command.

The following AWS CLI command shows how you can pass the JSON configuration within the CLI command, where you need to use the `--auto-scaling-role` parameter to specify an IAM role that has permissions to add and terminate instances and the `AutoScalingPolicy` option in the `--instance-groups` parameter to specify custom scaling rules:

```
aws emr create-cluster --release-label emr-6.3.0
--service-role EMR_DefaultRole --ec2-attributes
InstanceProfile=EMR_EC2_DefaultRole --auto-scaling-
role EMR_AutoScaling_DefaultRole  --instance-groups
Name=<MyMasterIG>,InstanceGroupType=MASTER,InstanceType=m5.
2xlarge,InstanceCount=1
'Name=<MyCoreIG>,InstanceGroupType=CORE,InstanceType=m5.
2xlarge,InstanceCount=2,AutoScalingPolicy={Constr
aints={MinCapacity=3,MaxCapacity=5},Rules=[{Name=
Default-scale-out,Description=Integrate scale-out
rule,Action={SimpleScalingPolicyConfiguration={AdjustmentType=
CHANGE_IN_CAPACITY,ScalingAdjustment=1,CoolDown=300}},Trigger=
{CloudWatchAlarmDefinition={ComparisonOperator=LESS_
THAN,EvaluationPeriods=1,MetricName=
YARNMemoryAvailablePercentage,Namespace=AWS/
ElasticMapReduce,Period=300,Statistic=AVERAGE,Threshold=15,
Unit=PERCENT,Dimensions=[{Key=JobFlowId,Value="${emr.
clusterId}"}]}}}]}'
```

Please replace `<MyMasterIG>` and `<MyCoreIG>` with the appropriate values before executing the command.

If you need to pass the configuration using a JSON file path, then you can refer to the following AWS CLI command, which assumes the configuration filename is `MyInstanceGroupConfig.json`:

```
aws emr create-cluster --release-label emr-6.3.0 --service-
role EMR_DefaultRole --ec2-attributes InstanceProfile=EMR_
EC2_DefaultRole --instance-groups  --auto-scaling-role EMR_
AutoScaling_DefaultRole
```

Please replace the `<path>` variable with your path before executing the command.

If you need to remove the autoscaling policy from an existing cluster, then you can execute the following command:

```
aws emr remove-auto-scaling-policy --cluster-id <Cluster-ID>
--instance-group-id <InstanceGroup-ID>
```

Please replace the `<Cluster-ID>` and `<InstanceGroup-ID>` variables with the appropriate values before executing the command.

In this section, we have dived deep into both EMR-managed scaling and autoscaling. Now let's understand how you can manually resize your cluster.

Manually resizing your EMR cluster

You have learned how you can integrate autoscaling into your cluster, but apart from autoscaling, you do also have the option to manually review your workloads and trigger manual resize action. You can trigger resize requests using the EMR console, the AWS CLI, and the EMR API.

Resizing a cluster using the AWS console

Refer to the following steps if you would like to resize your cluster instances using the EMR console:

1. After signing in to the AWS console, navigate to the Amazon EMR console at `https://console.aws.amazon.com/elasticmapreduce/`.

2. Choose the **Clusters** option and then select the active cluster that you would like to resize.

3. On the cluster detail page, navigate to the **Hardware** tab.

4. If you have configured instance groups for your cluster, then select **Resize** against the respective instance group, specify the new count, and finally, confirm by selecting the green checkmark.

    Alternatively, instead of an instance group, if you have configured an instance fleet for your cluster, then select **Resize** for the **Provisioned capacity** column. Specify new values for the on-demand units and spot units and click **Resize** to confirm.

When you resize the nodes, the instance group status changes and after completion, it goes into the **Running** state again:

Figure 6.8 – Screenshot of the EMR console showing the Resize option for an instance group

Now let's learn how to resize a cluster using the AWS CLI.

Resizing a cluster using the AWS CLI

You can resize a running cluster using AWS CLI commands, where you can increase core nodes and increase or decrease task nodes.

Assume your task instance group has five nodes and you need to increase it to seven; then, you can execute the following command, which uses the `modify-instance-group` option:

```
aws emr modify-instance-groups --instance-groups
InstanceGroupId=<instance-group-id>,InstanceCount=7
```

You can get the `<instance-group-id>` value by executing the following `describe-cluster` command. Please replace `<Cluster-ID>` with your EMR cluster ID:

```
aws emr describe-cluster --cluster-id <Cluster-ID>
```

If you need to terminate a specific instance of an instance group, you can execute the following command, which needs the specific instance ID, which you can get by executing the `aws emr list-instances --cluster-id <Cluster-ID>` command:

```
aws emr modify-instance-groups --instance-groups
InstanceGroupId=<ig-id>,EC2InstanceIdsToTerminate=<instance-id>
```

In this section, we have dived deep into different scaling aspects of EMR, specifically the autoscaling features such as managed scaling and autoscaling. Now let's understand how they compare to each other.

Comparing managed scaling with autoscaling

After learning about both EMR-managed scaling and autoscaling with a custom policy, let's compare both side by side to understand their differences and for which use case you can choose which one.

The following table draws a comparison between both of them:

	EMR-Managed Scaling	**Custom Autoscaling Policy**
Minimum Amazon EMR release version	5.30 and later.	4.0 and later.
Cluster types supported	Instance groups and instance fleets.	Instance groups only.
Scaling rules management	Amazon EMR-managed algorithm that constantly monitors different key cluster- and job-level metrics to make scaling decisions that optimize cluster resource utilization.	You can define custom scaling rules with different metrics that scale your cluster up or down with a fixed number of instances.
Configuration granularity	Cluster-level minimum or maximum constraints.	Instance group-level configuration.

	EMR-Managed Scaling	Custom Autoscaling Policy
Metric collection frequency to make scaling decisions	Every 1 to 5 seconds.	Every 5 minutes.
Evaluation frequency	Every 5 to 10 seconds.	Every 5 minutes.
Scaling algorithm	No configuration required. EMR's dynamic scaling algorithm takes care of it.	Allows you to configure a fixed number of instances for scaling up and down, when a defined condition is met.
Cooldowns between resizes	No configuration required. EMR's dynamic scaling algorithm takes care of it.	Allows you to define your custom cooldown periods between resizes.
Scaling based on custom metrics	No configuration required. EMR's dynamic scaling algorithm takes care of it.	Allows you to choose the metrics and their thresholds based on which you can do scaling operations.

Table 6.2 – A table showing a comparison between EMR-managed scaling and autoscaling with a custom policy

As you can see, EMR-managed scaling automates most of the scaling decisions and is great for YARN-based applications, whereas autoscaling with a custom policy is great when you want to have tighter control over the scaling rules.

Cluster cloning and high availability with multiple master nodes

You have learned about different cluster configurations, such as cluster scaling, debugging, and monitoring. Next, we will look at how to configure your EMR cluster to be highly available with multiple master nodes and how to clone an existing cluster that might be active or terminated.

High availability with multiple master nodes

Starting from EMR 5.23.0, you can launch an EMR cluster with multiple master nodes, which provides high availability for cluster applications such as YARN, HDFS NameNode, Spark, Hive, and Ganglia. You can use the EMR console or the AWS CLI to launch a cluster that has either one or three master nodes. If your cluster's primary master node fails or your NameNode or ResourceManager crashes, then EMR will automatically failover to stand by the master node, which makes the cluster fault-tolerant.

EMR automatically replaces the failed node with a new master node that has the same configuration and bootstrap actions as the failed master node.

To improve cluster availability, EMR can also take advantage of the EC2 placement groups to make sure master nodes are deployed on distinct underlying hardware. But the cluster and its nodes can only be placed in a single Availability Zone or subnet.

> **Important Note**
>
> As a prerequisite, please attach the `AmazonElasticMapReducePlacementGroupPolicy` AWS-managed policy to the EMR service role that you plan to use for launching your cluster.
>
> Also note that EMR automatically enables termination protection for clusters with multiple masters and you cannot enable auto-termination after a cluster is launched. If you need to terminate the cluster, then you need to disable termination protection first and then trigger a termination request.

Now, let's get an overview of the big data applications configured for high availability with a multi-master node cluster and how they work.

Applications supported with a multi-master node cluster

Not all the applications of the EMR cluster support high availability with multiple master nodes; the following are the ones supported at the time of writing this book, where each application's behavior varies depending on how it is integrated into EMR:

- **HDFS**: Out of the three master nodes, NameNode runs only on two master nodes where one acts as active and the other is on standby. When the primary master node or active NameNode fails, EMR automatically fails over to the standby master node and the standby NameNode becomes active to take over all operations of the cluster. After EMR replaces the failed master node, then it joins as the standby node.

  You can SSH to any of the master nodes and execute the following command to find which the active NameNode is:

  ```
  hdfs haadmin -getAllServiceState
  ```

- **YARN ResourceManager**: This runs in all three master nodes with one as active and the other two on standby. In the case of primary master node failure, the standby master node's **ResourceManager** becomes active and takes control of all the operations.

  You can access the `http://<master-public-dns-name>:8088/cluster` URL by replacing `<master-public-dns-name>` with any of the master nodes' public DNS names and it will automatically direct you to the active `ResourceManager`. If you need to identify the active `ResourceManager` using SSH, then execute the following command in any of the master node's shell prompts:

  ```
  yarn rmadmin -getAllServiceState
  ```

- **HBase**: This automatically fails over to stand by the master node and if you are connecting to HBase using REST or the Thrift server, then you must switch back to the new master node.

- **HCatalog**: Availability not affected as it's built on Hive Metastore, which exists outside of the cluster.

- **Spark**: Spark applications get executed in YARN containers and can also react to master node failover.

- **Sqoop**: You can configure Sqoop to store its metadata information in an external database to be highly available.

- **Tez**: This also runs in YARN, so is highly available.

- **Phoenix**: This runs in all three master nodes and its QueryServer runs in one of the master nodes. You can use the `/etc/phoenix/conf/phoenix-env.sh` file to find the private IP address of Phoenix's QueryServer.

- **JupyterHub**: Highly available as it is available in all three master nodes. But it's recommended to configure notebook persistence with Amazon S3 to prevent loss.

- **Zeppelin**: This is installed in all three master nodes and it stores its interpreter configuration and notes in HDFS by default to avoid data loss. Interpreter sessions are stored in master nodes, so they get lost in case of failures.

- **ZooKeeper**: This is highly available as it is the foundation for HDFS automatic failover.

- **Livy**: This is highly available as it is installed on all three master nodes and you can create a new session with the new master node.

- **Flink**: Its availability is not affected as its `JobManagers` run as YARN `ApplicationMaster` in core nodes.

- **Ganglia**: This is available in all master nodes, so its availability is not affected by the failure of an active master node.

- **Mahout**, **MXNet**, **TensorFlow**, and **Pig**: Their availability is not affected as they don't have any daemons.

> **Important Note**
> To configure Hive, Hue, PrestoDB, PrestoSQL, and Oozie as highly available with multiple master nodes, you should externalize their Metastore database so that their availability is not affected by master node failure.

Launching and terminating an EMR cluster with multiple master nodes

In the EMR console's advanced cluster creation option, you can enable multiple masters under the **Multiple master nodes (optional)** section of the **Software and steps** screen. Now let's understand how you can configure multiple masters while creating a cluster using the AWS CLI.

The following example AWS CLI command represents creating a cluster with multiple masters with a default AMI:

```
aws emr create-cluster --name "multi-master-
cluster" --release-label emr-6.3.0 --instance-groups
InstanceGroupType=MASTER,InstanceCount=3,InstanceType=m5.xlarge
InstanceGroupType=CORE,InstanceCount=5,InstanceType=m5.xlarge
--ec2-attributes KeyName=ec2_key_pair_name,InstanceProfile=EMR_
EC2_DefaultRole,SubnetId=<subnet-id> --service-role EMR_
DefaultRole --applications Name=Hadoop Name=Spark
```

Please replace `<subnet-id>` with your subnet ID. You can launch your cluster in both public and private subnets.

To terminate a cluster that has multiple masters, you need to disable termination protection first and then terminate the cluster. The following is an example of AWS CLI commands to terminate the cluster:

```
aws emr modify-cluster-attributes --cluster-id <Cluster-ID>
--no-termination-protected
aws emr terminate-clusters --cluster-id <Cluster-ID>
```

Please replace `<Cluster-ID>` with your cluster ID.

Considerations and limitations of a multi-master node cluster

The following are some of the considerations you should take note of when you are configuring your cluster with multiple master nodes:

- If you have connected to the active master node using SSH, then the connection will break in case of node failure. You can connect to the node again after EMR replaces it and note that the new node's public IP address will be different but its private IP address will be the same as the previously failed node.

- The Hive Metastore daemon runs in all master nodes, so in the case of primary master node failure, your application's **Java Database Connectivity (JDBC)** or **Open Database Connectivity (ODBC)**, connectivity will get terminated and you can connect to other active master nodes.

- Take a note of the following considerations for EMR steps:

  - In the case of master node failure, all the steps running on the master node will be marked as FAILED and its local data will be lost, but you should check the output of the step to reflect the real state of the step.

  - If a step has started as a YARN application and is running when the master node fails, then because of automatic failover of the master node, it will continue and succeed.

  - To let the cluster continue and allow failover of the master node, it's recommended that you set the ActionOnFailure parameter to CONTINUE or CANCEL_AND_WAIT instead of TERMINATE_JOB_FLOW or TERMINATE_CLUSTER.

- To use Kerberos authentication in your EMR cluster, you have to configure an external **Key Distribution Center (KDC)**.

- If your cluster's subnet is oversubscribed or fully utilized, then in case of failure, EMR cannot replace your failed master node. To avoid such a scenario, it's recommended that you assign an entire subnet to your EMR cluster and make sure it has enough private IP addresses.

Apart from these considerations, the following are a few of the limitations that you should also take note of:

- As a limitation, **instance fleets**, **EMR notebooks**, **persistent application user interfaces**, and **one-click access** to persistent Spark history server features are not available in EMR clusters with multiple master nodes.

- If two master nodes fail at the same point in time or the whole Availability Zone goes down, then your EMR cluster cannot recover.

For a detailed list of limitations, please refer to the AWS documentation.

Cloning an existing EMR cluster

In the EMR console, you have the option to clone an existing cluster irrespective of its current state and optionally, the cloned cluster can include an existing cluster's steps too.

The following steps explain how you can clone a cluster:

1. Navigate to the Amazon EMR console.
2. Select **Clusters**, which lists all clusters, including clusters that are in a terminated state.
3. Select the cluster that you want to clone and click **Clone**.
4. There will be a popup that will ask **Would you like to include steps?**. Select **Yes** if you need to and then click **Clone**.
5. This will open up EMR's advanced cluster creation screens with all configurations populated that match the existing cluster, which you planned to clone. Review the configurations and click **Create cluster**.

This should create the new EMR cluster, which will have the existing cluster's configuration and steps. If you are using AMI release 3.1.1 (Hadoop 2.x), AMI release 2.4.8 (Hadoop 1.x), or anything later than that, then you can clone up to 1,000 steps, but if you are using earlier releases of AMI, then you can clone a maximum of 256 steps.

> **Important Note**
> It is highly recommended to configure both scale-up and scale-down policies when you configure autoscaling with a custom policy, which will optimize your cluster's resource utilization and can provide better performance and cost savings. Defining either scale out or scale in without the other one will require you to take manual action.

Summary

Over the course of this chapter, we got an overview of how to monitor cluster and job activities using a cluster's application interfaces, cluster metrics, and the CloudWatch console. We also saw how to enable auditing on cluster API activities using AWS CloudTrail.

Then, we dived deep into EMR cluster scaling capabilities, which includes EMR-managed scaling and autoscaling with custom policies. We also learned how they compare to each other.

Finally, we covered how to make our cluster highly scalable with multiple master nodes and what the supported applications are. We also learned how we can clone an existing cluster to replicate its configurations and steps.

That concludes this chapter! Hopefully, you got a good overview of monitoring, scaling, and high-availability aspects of the cluster, and in the next chapter, we can dive deep into security aspects of EMR.

Test your knowledge

Before moving on to the next chapter, test your knowledge with the following questions:

1. Assume you have a long-running EMR cluster that is being used by multiple teams for ETL jobs and data analysis. Because of its multi-tenant nature, your organization asks that you provide a report of who is accessing the cluster and for which activities. How would you prepare such a report and from where will you collect this information?

2. Assume you have a long-running EMR cluster that integrates instance fleets into its configurations. Your cluster has one master and three core nodes to start with and you are planning to benefit from EMR scaling capabilities so that when you have more workload, your cluster will scale up, and when the jobs are finished, it will scale down. Out of EMR-managed scaling and autoscaling with custom policies, which one will you choose?

3. You have a long-running EMR cluster that is being used by multiple teams of your organization. You have configured Hive, Hue, and Spark applications on your cluster and are using Amazon S3 as the cluster persistent storage layer. Your users are using the Hue interface to execute Hive queries. You are expected to make this setup fault-tolerant so that your Hue users don't lose access to the cluster or don't have downtime to execute Hive queries. How would you set up your cluster?

Further reading

The following are a few resources you can refer to for further reading:

- Learn how to configure the FoxyProxy or SwitchOmega plugins for the SOCKS proxy: `https://docs.aws.amazon.com/emr/latest/ManagementGuide/emr-connect-master-node-proxy.html`.

- EMR cluster step events with Amazon CloudWatch: `https://docs.aws.amazon.com/emr/latest/ManagementGuide/emr-manage-cloudwatch-events.html`.

- Configuring CloudWatch event rules: `https://docs.aws.amazon.com/AmazonCloudWatch/latest/events/Create-CloudWatch-Events-Rule.html`.

- EMR integration with EC2 placement groups: `https://docs.aws.amazon.com/emr/latest/ManagementGuide/emr-plan-ha-placementgroup.html`.

7
Understanding Security in Amazon EMR

In the previous chapter, you learned about EMR cluster monitoring, scaling, high availability, and cloning capabilities.

When you implement solutions in AWS, security is the most important thing that you should be focusing on. These security aspects include infrastructure security, network security, and data-level security. AWS provides several services and features by means of which you can implement security around your solution.

In this chapter, we will explain how you can control authentication and authorization in relation to your cluster, how you can secure data with encryption at rest and in transit, and finally, how AWS IAM, VPC, subnets, and cluster security groups play a role in making the cluster secure.

Now, let's dive deep into the following topics and understand how they help in implementing security in Amazon EMR:

- Understanding the basics of security
- AWS IAM integration with Amazon EMR
- Understanding data protection in EMR
- Role of security groups and interface VPC endpoints

Technical requirements

In this chapter, we will dive deep into the different security aspects of EMR, including IAM access permissions, data encryption, and controlling network traffic to the EMR cluster. Before getting started, please make sure you have access to the following resources.

- An AWS account
- An IAM user who has permission to create and manage an EMR cluster with related resources, including Amazon EC2 instances, required IAM roles, and security groups
- IAM access privileges to create VPC endpoints as well as create and manage encryption keys using AWS KMS
- Access to EMR security documentation that is available through `https://docs.aws.amazon.com/emr/latest/ManagementGuide/emr-security.html`

Now, let's understand what it means when we talk about security in EMR and how the shared responsibility model works.

Understanding the basics of security

AWS has always given top priority to security and highlighted security as the most important aspect to consider when getting started. When we talk about security, you need to understand that it's a shared responsibility between AWS and its customers. AWS provides the infrastructure and security features by means of which you can implement security, and it is your responsibility to implement security as per your requirements.

The following diagram represents what is included within the sphere of responsibility of both AWS and the customer:

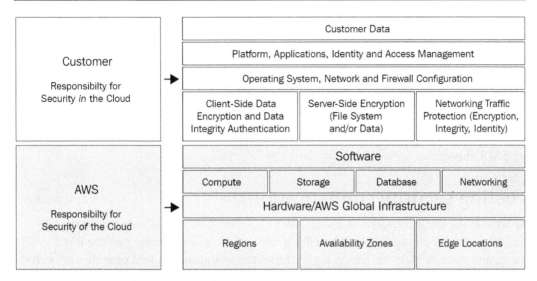

Figure 7.1 – Diagram representing a shared responsibility model for AWS

At a high level, you can divide security-shared responsibility into the following categories:

- **AWS responsibility for security of the cloud**: AWS provides infrastructure and related services by means of which customers can build and deploy their applications to the AWS public cloud. AWS infrastructure comprises the hardware, networking, software, and data centers or facilities that run AWS Cloud services. AWS is responsible for protecting or securing these infrastructure components, which run all the services offered by AWS.

- **Customer responsibility for security in the cloud**: You, as a customer, are responsible for integrating security features when you start using any AWS services. Depending on the AWS services you choose for your implementation, you will have different configurations as part of your scope to implement security; for example, Amazon EC2 categorized as **Infrastructure as a Service (IaaS)**, which requires you or your customers to perform operating system updates, apply security patches, configure SSH or public or private IP-based access, placing them under VPC, subnet and security groups, and managing and owning the software application installed on them. For serverless services, however, such as AWS Lambda or DynamoDB, AWS owns the infrastructure and you are responsible for your data and controlling access to it.

When we dive deep into EMR security, we can understand what is within your or your customers' scope as part of the shared responsibility model.

Amazon EMR provides security configurations as a feature through which you can define authentication, authorization, encryption, and other configuration once and then attach them to multiple clusters. Starting with the EMR 4.8.0 release, you can define data encryption settings using security configurations. However, security configuration for Kerberos authentication and Amazon S3 authorization for EMRFS is only available starting with the EMR 5.10.0 release.

Next, let's learn how you can create a security configuration and how you can specify it for your EMR cluster.

Creating security configurations

You can use the EMR console, AWS CLI, AWS SDKs, or AWS CloudFormation template to create a security configuration. We will explain how you can leverage just the EMR console and the AWS CLI for the configuration settings and that should provide you with a starting point for building more dynamic applications using AWS SDK or for building DevOps automation using AWS CloudFormation templates.

Creating a security configuration using the EMR console

To create a security configuration using the EMR console, you can refer to the following steps:

1. Navigate to the EMR console at `https://console.aws.amazon.com/elasticmapreduce/`.

2. In the navigation pane, under **EMR on EC2**, choose **Security Configurations** and then click the **Create** button, which will open up a new screen where you can define configurations.

3. Specify a name for the security configuration.

4. Choose the **Encryption**, **Authentication**, **Authorization**, and **EC2 Instance Metadata Service** options and then click **Create**.

This will create a security configuration that you can apply to an EMR on the EC2 cluster.

Creating a security configuration using the AWS CLI

To create a security configuration, you can SSH to any of the master nodes and execute the following command, which employs the `create-security-configuration` option:

```
aws emr create-security-configuration --name "<SecConfigName>"
--security-configuration <MySecConfig-FilePath>
```

Please replace `<SecConfigName>` with your security configuration name. Then, store the configuration options in a JSON file and replace `<MySecConfig-FilePath>` with the JSON configuration file path. The path might look like this – `file://<MySecConfig>.json`.

Specifying a security configuration for your cluster

After you have created the security configurations, as a next step, you can assign that security configuration to one or more EMR clusters using the EMR console or AWS CLI. Let's understand how you can specify the configuration for your cluster.

Specifying a security configuration using the EMR console

As explained in *Chapter 5*, *Setting Up and Configuring EMR Clusters*, when using the AWS console to create an EMR on an EC2 cluster, you can specify the security configuration during **Step 4: Security** of the advanced cluster create option. The following is a screenshot of the EMR console that represents the same:

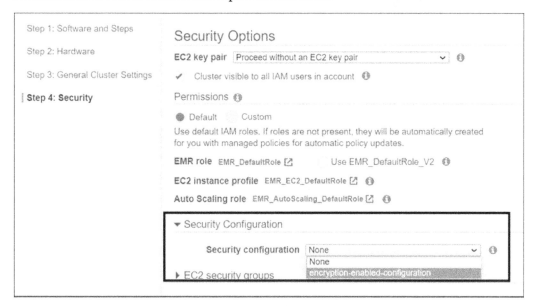

Figure 7.2 – EMR console for specifying security configurations

You need to choose the security configuration from the drop-down option. This will list all the security configurations you have already created.

Specifying a security configuration using the AWS CLI

Before applying a security configuration using the AWS CLI, you need to make sure you have selected EMR release 4.8.0 or later. As represented in the following AWS CLI command sample, you can specify the security configuration name using the `--security-configuration` parameter:

```
aws emr create-cluster --instance-type m5.2xlarge --release-
label emr-6.4.0 --security-configuration <mySecurityConfigName>
```

Before executing the command, please replace the `<mySecurityConfigName>` variable with the name of your security configuration.

This section provided you with an overview of how you can create a security configuration and how you can assign it to your cluster. Next, let's dive deep into each of the security topics.

AWS IAM integration with Amazon EMR

The AWS **Identity and Access Management** (**IAM**) service assists in integrating authentication and authorization mechanisms on top of AWS services or APIs. You can use IAM users, groups, or roles to define permission policies.

In Amazon EMR, using IAM identity-based policies, you can define which IAM user, group, or role can access which specific resources and, on a specific resource, which actions are allowed or denied. You can also specify conditions on which basis a specific action on a resource is allowed, or not. Please note that Amazon EMR does not support resource-based policies.

The following are the three primary components of an IAM policy:

- **Actions**: Policy actions specify which action on your EMR cluster is allowed or denied and uses the `elasticmapreduce:` prefix before the action. For example, a described cluster action will have `elasticmapreduce:Describe` as an action. Your policy statements define either an `Action` or `NotAction` parameter. A `NotAction` parameter means to allow all actions except a specific one.

   The following example shows how you can specify multiple actions with comma-separated values:

   ```
   "Action": [
           "elasticmapreduce:Describe",
           "elasticmapreduce:RunJobFlow"
   ```

You can also specify multiple actions with wildcard (*) syntax, for example, `elasticmapreduce:Describe*`.

- **Resources**: The `Resource` element of the policy defines the object or EMR cluster to which the action is applicable. The value of the resource is either your EMR cluster's ARN or it can be a wildcard (*) to specify all clusters.

- **Condition Keys**: This is an optional element of the policy that allows you to specify conditions to match the policy to be in effect. The matching conditions can be a string match with a value or with less than or greater than operators.

> **Important Note**
>
> For multiple conditional statements, AWS IAM evaluates them with the logical AND operator, whereas if multiple values are specified for a single condition, then IAM evaluates them with the logical OR operator.

After understanding the components of an IAM policy, let's learn how you can configure service roles for your EMR cluster.

Configuring an IAM service role for your EMR cluster

An EMR cluster needs different service roles for different operations. Service roles are the ones that provide permissions to EMR applications such as Hadoop, Hive, and Spark to access other AWS services and perform different actions for their execution. Each EMR cluster must have a service role for Amazon EMR to help perform service-level operations and a service role for an EMR cluster's EC2 instances to interact with other AWS services. Apart from this, an EMR cluster also needs a service role to scale cluster resources if you have configured autoscaling on your cluster and a service role for your EMR notebook, if you have set up an EMR notebook.

As highlighted earlier, EMR provides default IAM roles for your cluster which will have default managed policies integrated into it. Managed IAM policies are managed by AWS and are updated automatically, as required by the service. Even if EMR provides default roles, you can create your own IAM roles and assign them to your cluster while creating them.

Service roles used by EMR

The following are the service roles used by EMR for which EMR creates default roles with AWS managed policies, but it lets you create and assign any custom roles you have created for them:

- Service role for Amazon EMR (EMR role)
- Service role for EMR cluster EC2 instances (EC2 instance profile)
- Service role for automatic scaling in EMR (Autoscaling role)
- Service role for EMR notebooks
- Service-linked role

Now, let's dive deep into each of these roles to understand their functions better and ways to implement custom roles for them.

EMR role

An EMR service role includes permissions or IAM policies that allow EMR to provision cluster resources such as provisioning EC2 instances while creating a cluster, and performing service-level tasks, such as EMR applications interacting with other AWS services.

By default, EMR creates `EMR_DefaultRole`, which includes `AmazonEMRServicePolicy_v2` managed policy. Since managed policies are created and maintained by AWS, they are subject to change as required by the service role. Please note that in order to use this managed policy, you will need to pass `for-use-with-amazon-emr-managed-policies = true` in the policy condition, as shown in the following:

```
"Condition": {
        "StringEquals": {
            "aws:ResourceTag/for-use-with-amazon-emr-managed-policies": "true"
        }
    }
```

The preceding code snippet is an example that shows the condition section of the managed policy.

The EC2 instance profile service role is assigned or attached to each of the EC2 instances launched as part of the EMR cluster. Application processes that run on top of Hadoop in these EC2 instances or nodes assume that this EC2 instance profile role interacts with other AWS services.

The default role created by EMR for this is `EMR_EC2_DefaultRole`, which uses the `AmazonElasticMapReduceforEC2Role` policy.

> **Important Note**
>
> The `AmazonElasticMapReduceforEC2Role` managed policy will be deprecated and the EMR service will not replace it with any other default policy. It is recommended that instead of using the default managed policy, you should apply resource-based policies to Amazon S3 buckets and other resources or create your own custom policy and role to use as an instance profile role.

Your EC2 instance profile role will require the following additional permissions:

- Reading and writing to Amazon S3 using EMRFS: Amazon EMR uses the service role for the cluster EC2 instances to interact with the Amazon S3 path. So, to provide access, you need to provide specific bucket and folder permissions to your EC2 instance profile role.

- If you have enabled the archiving of log files to Amazon S3, then you must specify the `s3:PutObject` action permission on your log directory S3 resource path.

- If you have enabled debugging on your cluster, then you need to provide access to the `sqs:GetQueueUrl` and `sqs:SendMessage` actions on your specific SQS queue.

- If you are going to use AWS Glue Catalog integration for your EMR cluster, then you should be specifying `glue:` prefix actions so that EMR service applications can interact with AWS Glue Catalog databases and tables.

Please refer to the AWS documentation for a detailed policy document that you should be attaching to your role.

Autoscaling role

This role is required by Amazon EMR to scale your cluster up and down, where it looks at CloudWatch and cluster metrics then takes scaling decisions to add or terminate EC2 instances.

The following shows the permissions included in the managed policy that might change in the future as EMR service requirements change:

```
{
    "Version": "2012-10-17",
    "Statement": [
        {
            "Action": [
                "cloudwatch:DescribeAlarms",
                "elasticmapreduce:ListInstanceGroups",
                "elasticmapreduce:ModifyInstanceGroups"
            ],
            "Effect": "Allow",
            "Resource": "*"
        }
    ]
}
```

The default role that EMR creates is `EMR_AutoScaling_DefaultRole`, which includes an `AmazonElasticMapReduceforAutoScalingRole` managed policy.

EMR notebook role

When you create an EMR notebook, it needs privileges to interact with other EMR applications and other AWS services. The default role that EMR creates for this is `EMR_Notebooks_DefaultRole`, which includes `AmazonElasticMapReduceEditorsRole` and `S3FullAccessPolicy` managed policies by default.

If you are creating your custom role for your notebook, make sure you provide at least the following S3 privileges to access your S3 resources:

```
"s3:PutObject",
"s3:GetObject",
"s3:GetEncryptionConfiguration",
"s3:ListBucket",
"s3:DeleteObject"
```

In addition, if your Amazon S3 buckets are encrypted, then you need to provide the following additional privileges:

```
"kms:Decrypt",
"kms:GenerateDataKey",
"kms:ReEncrypt",
"kms:DescribeKey"
```

If you are going to link GitHub repositories to your EMR notebook and are planning to enable encryption, then you must provide the `secretsmanager:GetSecretValue` privilege to your EMR notebook service role.

Service-linked role

This service role is used by EMR to clean up EC2 resources when they are no longer in use. This role works along with your cluster's EC2 instance profile role and EMR role to trigger EC2 actions.

The default role that EMR created for this is `AWSServiceRoleForEMRCleanup` and it has the trust policy defined for the EMR service, which is `elasticmapreduce.amazonaws.com`.

Unless defined otherwise, this role can only be assumed by EMR. This role is automatically created by EMR when you launch an EMR cluster if it does not exist. You can plan to delete this role once you have deleted all the EMR clusters in your account.

To delete this role, you can refer to the following steps in the EMR console and make sure you have terminated all your EMR clusters before doing this:

1. Navigate to the AWS IAM console at `https://console.aws.amazon.com/iam/`.

2. Click **Roles** from the navigation pane and then click the `AWSServiceRoleForEMRCleanup` role name.

3. On the role's summary screen, click the **Access Advisor** tab and review recent activity to confirm whether the role is being used by any EMR cluster. If it is being used, then deletion will fail.

4. Next, you can navigate back to the role list and select the checkbox beside the `AWSServiceRoleForEMRCleanup` role.

5. Then, click **Delete** and confirm the action from the confirmation dialog.

The deletion process is asynchronous, so it might succeed or fail. If it fails, click **View details** or **View resources** from the IAM notifications to learn about the reason for failure and take the necessary action.

Configuring IAM roles for EMRFS

As described previously, EMR assumes the EC2 instance profile role if it needs to access Amazon S3 resources using EMRFS. The same service role is used to access S3 resources, regardless of which Amazon S3 path or user or group is requesting access.

In case you have an EMR cluster that has multiple IAM users that requires a different level of access, then you can set up configurations with the IAM role for EMRFS. EMRFS has the flexibility to assume different service roles based on the Amazon S3 path or the user or group requesting the access. You can configure each IAM role for EMRFS to have different permissions to access Amazon S3 data. The IAM role for EMRFS supports the EMR 5.10.0 release.

To configure IAM roles for EMRFS, you need to set up role mappings that specify IAM roles corresponding to identifiers who can be users, groups, or Amazon S3 prefixes. These identifiers in the mapping decide access to S3 using EMRFS. Once you do this, the following processes take place:

- When EMRFS makes a request to access S3 data, it matches the access request with the security configuration mapping defined in top-down order.

- If the request matches, then EMRFS assumes the respective IAM role to access S3 data instead of assuming the EC2 instance profile role.

- In case the request does not match any of the mappings, then by default, it assumes the EC2 instance profile role.

Please note that IAM roles for EMRFS do not provide any host-level isolation and are limited to application-level isolation that represents controlling which user can access which application.

Setting up security configurations with IAM roles for EMRFS

To set up the security configurations with IAM roles for EMRFS, you can use the EMR console or AWS CLI. As explained previously in the *Understanding the basics of security* section, it's a two-step process where first you create a security configuration and then assign it to your cluster.

The following is a screenshot of **IAM roles for EMRFS** configuration, available in the EMR console's create security configuration screen, where you specify IAM roles for EMRFS:

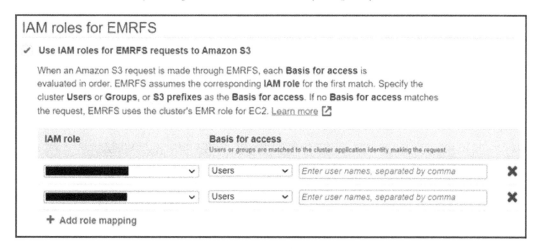

Figure 7.3 – EMR console security configuration to specify IAM roles for EMRFS

After you have created the security configuration, you can assign it to your cluster while creating it through the EMR console or AWS CLI.

Integrating IAM roles in applications that invoke AWS services directly

Applications that run on your cluster EC2 instances can use the EC2 instance profile to get temporary security credentials while calling other AWS services. Starting with the EMR 2.3.0 release, Hadoop applications available in your EMR cluster are already updated to leverage IAM roles.

If your application uses Hadoop architecture for its execution and does not directly invoke any AWS services, then it should work with IAM roles as they are updated to work without any modifications.

However, if your application calls AWS services directly, then you will have to update your application's way of working so that it does not get account credentials from the /etc/hadoop/conf/core-site.xml configuration file of the EC2 instance. Rather, it follows either of the following approaches to get temporary credentials:

- Your application uses AWS SDK, which uses IAM roles to get temporary credentials to access AWS services

- Alternatively, you can call the EC2 instance metadata URL from the EC2 instance to get the temporary security credentials. The following is an example command that shows how you can call the EC2 instance metadata URL and it uses the default EC2 instance profile, `EMR_EC2_DefaultRole`:

```
GET http://169.254.169.254/latest/meta-data/iam/
security-credentials/EMR_EC2_DefaultRole
```

This command returns the `AccessKeyId`, `SecretAccessKey`, `SessionToken`, and `Expiration` attributes.

Please refer to the AWS documentation to refer to the programming language SDKs by means of which you can get the temporary credentials.

Allowing users and groups to create and modify roles

To create and/or modify EMR clusters or to create custom roles that will be attached to the EMR cluster that overrides the default roles, the IAM users or groups must have the following IAM action permissions:

- `iam:CreateRole`
- `iam:PutRolePolicy`
- `iam:CreateInstanceProfile`
- `iam:AddRoleToInstanceProfile`
- `iam:ListRoles`
- `iam:GetPolicy`
- `iam:GetInstanceProfile`
- `iam:GetPolicyVersion`
- `iam:AttachRolePolicy`
- `iam:PassRole`

Out of all these IAM actions, `iam:PassRole` is needed to create a cluster, while the remainder are all required to create or modify IAM roles.

Identity-based policies and best practices

By default, none of the IAM users or roles are permitted to create or modify Amazon EMR resources using the EMR console, AWS CLI, or AWS APIs. It is the IAM administrator who needs to create IAM policies and assign them to specific users, groups, or roles to perform specific API operations or actions.

An IAM policy that is attached to an IAM identity is called identity-based policy. It includes AWS-managed policies, custom policies, or inline policies that you embed in a specific user, group, or role.

As a best practice, it is recommended to use custom-managed policies instead of inline policies as inline policies cannot be reused. While configuring identity-based policies, you should consider the following general best practices:

- **Getting started with AWS managed policies**: To get started with EMR, you should go with default roles or managed policies already available in AWS IAM. This way, you can avoid the custom policy creation and can validate your setup quickly.

- **Granting least privileges**: This is a general security best practice across AWS services, where you create your custom role with the least privileges and add additional permissions as required. This will provide you with better granular security control.

- **Enabling Multi Factor Authentication (MFA)**: For additional security, you should consider enabling MFA for sensitive API operations.

- **Using policy conditions for additional granular security**: This allows you to specify conditions on which basis your defined action will be allowed. As an example, you can specify a set of IP addresses from which the request will be allowed.

Apart from IAM-based permissions, you should also tighten your security group access to make sure only allowed source addresses can access your EMR cluster nodes or user interfaces.

Understanding authentication to cluster nodes

As explained in *Chapter 6, Monitoring, Scaling, and High Availability*, you can SSH to cluster nodes using the EC2 key pair that you have specified during cluster setup. You can also configure SSH tunneling to the cluster master node with proxy setup to access the EMR application's web interfaces.

Starting with the EMR 5.10.0 release, you can also configure Kerberos to authenticate user access using your corporate credentials.

After understanding how IAM roles are integrated and configured in the EMR cluster, in the next section, we will learn how you can protect data at rest or in transit.

Understanding data protection in EMR

As you have learned in relation to the shared responsibility model in the *Understanding the basics of security* section of this chapter, you are responsible for maintaining the security of your applications and data by integrating security configurations and controls provided by AWS. As part of the security implementation, you can make your data secure, both in transit and at rest.

The following are some of the high-level security guidelines that you can follow to make your data secure:

- Follow data governance practices to define your user personas and the access privileges they will have.

- Define IAM users, groups, and roles as per the user personas and application requirements and observe the guidelines regarding least privilege.

- Use MFA for additional security on sensitive accounts or data access.

- Leverage the TLS protocol to communicate with AWS services or resources.

- Leverage AWS CloudTrail logs for auditing user actions on different APIs.

- Integrate data encryption solutions for data at rest.

- Take advantage of AWS services such as **Amazon Macie** to detect **Personally Identifiable Information** (**PII**) data in Amazon S3 and take action if you need to mask them before exposing them to data consumers.

- Avoid storing sensitive information in free form text fields.

While discussing the different security guidelines, encryption plays an important role in making your data secure at rest or in transit. Starting with the EMR 4.8.0 release, you can specify data encryptions for your EMR cluster using security configuration settings. Using security configurations, you can enable encryption for data in transit and for data at rest when it is stored in Amazon EBS volumes or Amazon S3 using EMRFS.

Let's dive deep into the encryption options you have and how you can configure them.

Encrypting data at rest for EMRFS on Amazon S3 data

When you enable encryption for EMRFS with Amazon S3, then the encryption works while reading from and writing to Amazon S3. You can specify **Amazon S3 Server-Side Encryption** (**SSE**) or **Client-Side Encryption** (**CSE**) as your default encryption mode for encryption at rest, but you can also override the encryption method for any specific bucket or Amazon S3 paths.

Apart from encryption at rest, **Transport Layer Security (TLS)** is enabled for encrypting the data in transit between your EMR cluster and Amazon S3.

You can leverage **AWS Key Management Service (KMS)** to get custom encryption keys for integration, but it might add additional storage and AWS KMS usage costs.

Amazon S3 SSE

When you have enabled encryption for Amazon S3, the data is encrypted at the object level when it writes data to underlying storage and is decrypted when the data is accessed.

The following are two different encryption key management systems where Amazon S3 with EMR is supported:

- **SSE-S3**: SSE is the mechanism where Amazon S3 manages the key.
- **SSE-KMS**: You can use AWS KMS with the customer master key, which is set up with policies required by Amazon EMR.

Please note that Amazon S3 SSE with customer-provided keys (SSE-C) is not supported with Amazon EMR at the time of writing this book.

Amazon S3 CSE

When you integrate client-side encryption on your Amazon S3 datasets, the encryption and decryption happen on the client layer, which means in the EMRFS client of your EMR cluster. Objects are encrypted within your EMR cluster before getting written to the S3 prefix and get decrypted in EMR after they are downloaded from S3.

The encryption key is provided by you, which can be key managed by AWS KMS (CSE-KMS) or your custom Java class that generates the **client-side master key (CSE-C)**. Depending on which method you use to generate the encryption key and the metadata of the object being encrypted, the specifics of the encryptions differ.

> **Important Note**
> Amazon S3 CSE only applies to EMRFS with S3 and not the cluster disk level datasets. Also note that, at the time of writing this book, **Hue** does not make use of EMRFS, so any object that we write to Amazon S3 using the Hue web interface is not encrypted.

EMR cluster local disc encryption

As explained in *Chapter 2*, *Exploring the Architecture and Deployment Options*, the EMR cluster's local data can reside either in the cluster's HDFS storage or in each individual node's EBS or instance store volumes. When you think of encrypting cluster local data, then you need to consider all these three storage layers. Now, let's get an overview of the encryption mechanism applied to each one of them:

- **HDFS encryption**: During distributed processing, Hadoop applications exchange data between cluster instances using HDFS, which involves instance store and EBS volumes. During distributed processing, data is exchanged between cluster instances by HDFS. When you enable local disk encryption, you can consider the following open source Hadoop encryption options:

  - `Secure Hadoop RPC` is set to `Privacy` and uses the **Simple Authentication and Security Layer** (**SASL**).

  - Data encryption on HDFS block data transfer that is `dfs.encrypt.data.transfer` is set to `true` and is configured to use **Advanced Encryption Standard** (**AES**) 256 encryption.

- **Instance store encryption**: Regardless of EMR settings, Amazon EC2 instance types that use **Non-Volatile Memory Express** (**NVMe**)-based SSDs as their instance store volumes use NVMe encryption. If your EC2 instance does not use NVMe-based SSDs, then EMR uses **Linux Unified Key Setup** (**LUKS**) to encrypt the instance store volumes when you enable local disk encryption, irrespective of the EBS volume encryption methodology.

- **EBS volume encryption**: If you have created your EMR cluster in an AWS region that has enabled EBS volume encryption by default, then its contents are encrypted irrespective of the EMR cluster encryption settings. However, if you have local disk encryption enabled using EMR security configurations, then that takes precedence over your EC2 instance default encryption settings. The following options are available when you enable encryption using security configuration settings:

  - **EBS encryption**: Starting with the EMR 5.24.0 release, you have the option to enable EBS encryption that encrypts your EBS root volume as well as any attached storage volumes. Please note that it is only available when AWS KMS is integrated as the key provider and is a recommended option too.

  - **LUKS encryption**: If you select this setting for cluster EBS volumes, then it is only applicable to attached storage volumes and not the root volume. Similar to S3 CSE, you can leverage either AWS KMS or a custom Java class to be your key provider service.

It is recommended to use the `DescribeVolumes` API to confirm the status of EBS encryption (should be enabled) on your cluster, since running `lsblk` on the cluster will only provide the status of LUKS encryption. `lsblk` is a Linux command that reads system files to provide information about all the block devices attached to the system.

Encrypting data in transit for EMRFS on Amazon S3 data

For in-transit encryption, several options are available, and these are EMR application-specific. You can enable application-specific encryption features using EMR security configurations and the following are the features that you can activate:

- **Hadoop**: Hadoop MapReduce shuffle uses TLS for encryption. `Secure Hadoop RPC` is set to `Privacy` and uses the SASL when you have enabled encryption at rest. Also, when HDFS blocks get transferred, they use the **Advanced Encryption Standard (AES)** 256 protocol for encryption and it is also activated when you enable encryption at rest on your cluster.

- **Spark**: Starting with EMR release 5.9.0, Spark also uses the AES 256 cipher for encrypting internal RPC communication, such as a block transfer between Spark components or shuffling between nodes. For earlier EMR releases, Spark used SASL with DIGEST-MD5 for its internal communication.

- **Presto**: Starting with EMR release 5.6.0, Presto uses SSL/TLS for internal communication between Presto nodes.

- **Tez**: `Tez shuffle handler` uses TLS (`tez.runtime.ssl.enable`).

- **HBase:** When you have enabled Kerberos authentication on your cluster, HBase sets the `hbase.rpc.protection` property to `privacy` for its encrypted communication.

When you plan to implement encryption for data in transit, then you can specify the encryption artifacts either by uploading zipped certificate files to Amazon S3 or by referencing your custom Java class.

Role of security groups and interface VPC endpoints

In previous sections of the chapter, you have learned how you can control access to your cluster using IAM permissions and how you can make your data secure at rest or in transit. In this section, you will learn about controlling access to your cluster using cluster security groups and how you can use VPC interface endpoints.

Controlling cluster network traffic with security groups

Security groups in AWS act as firewalls for your cluster EC2 instances, where you can control both inbound and outbound traffic. For example, you can define inbound rules to allow only your IP address to be the source of the SSH connection to your cluster nodes and you can add multiple rules for different access requirements.

You have two types of security groups; one is **managed security groups** that is created and managed by EMR, and the other is **custom-managed security groups** that you can create and assign to your EMR cluster. The custom security groups are optional, which you can assign to your cluster, in addition to the managed security groups. They contain the rules you specify and are not modified by EMR.

The EMR managed security group has rules that allow the EMR cluster to interact with other AWS services. You can modify the managed security group rules but you need to be very careful doing that as any miss can block access to your cluster.

As a best practice when it comes to providing least privileges for your cluster, it is recommended to avoid providing public access to your cluster by allowing inbound traffic from sources as IPv4 `0.0.0.0/0` or IPv6 `::/0`.

Working with EMR managed security groups

When you launch a cluster, you can attach different managed security groups with both **Master** and **Core & Task** node types of the cluster. If you are launching a cluster within a private subnet of the VPC, then you will have to specify an additional **managed security group** for service access.

EMR automatically creates managed security groups if they don't exist and are then assigned to your cluster. You need to make sure to either integrate default managed security groups available in EMR or integrate custom security groups for your cluster because a combination of both is not supported.

The following are the default managed security groups that EMR creates:

- **ElasticMapReduce-master**: If you have selected a public subnet for your cluster, then EMR specifies `ElasticMapReduce-master` as the default managed security group for the master node.

- **ElasticMapReduce-slave**: For the EMR cluster in a public subnet, the default managed security group for **Core & Task** instances is `ElasticMapReduce-slave`.

- **ElasticMapReduce-Master-Private**: If you have selected a private subnet for your cluster, then EMR uses `ElasticMapReduce-Master-Private` as the default security group name for the master node.

- **ElasticMapReduce-Slave-Private**: For clusters in a private subnet, EMR uses `ElasticMapReduce-Slave-Private` as the default managed security group for **Core & Task** nodes for clusters in a private subnet.

- **ElasticMapReduce-ServiceAccess**: For service access in private subnets, EMR uses the default security group name as `ElasticMapReduce-ServiceAccess`. It has inbound and outbound rules integrated that allow traffic over HTTPS using port `8443` and port `9443` to the other managed security groups in the private subnets. It also enables the cluster manager to communicate with the master, core, and task nodes.

If you are creating custom security groups for your cluster, then you can refer to the preceding default security group's inbound and outbound rules to make sure you are not missing any default rules that are needed in order for your cluster to function as expected.

If you would like to allow SSH access to your master node from specific trusted sources, then you can edit the inbound rules of the default `ElasticMapReduce-master` security group and add `SSH` access over port `22` for a specific source IP or can select `My IP` as the source that will autopopulate your IP address.

Working with additional custom-managed security groups

You have learned in the previous section that in the case of **EMR managed security group** settings, you have the option to select either default-managed security groups or custom security groups. But apart from that, optionally, you can also attach an additional security group that has your custom rules that could be controlling access from your client applications or allowing communication between different EMR clusters.

For example, you have multiple EMR clusters that are launched in the same VPC and subnet. You need to allow SSH inbound access to the master node for just a specific subnet of the cluster. To implement this, on top of the default managed security groups, you can add an additional security group that allows SSH inbound access over port 22 from the master node security group to all the clusters in the same subnet.

As explained earlier, you can apply a maximum of up to four additional security groups for each of the node types, such as **Master** and **Core & Task**, and four in a private subnet for service access. Also note that the maximum number of security groups you can add is also dependent on any AWS account level limits you might have.

Specifying EMR-managed security groups and additional custom security groups for a cluster

If you do not specify security groups, EMR creates default security groups for you, and you can modify them or add additional custom security groups. You can specify security groups on your cluster using the EMR console, AWS CLI, or EMR API.

Let's understand how you can specify the security groups on your cluster.

Specifying a security group using the EMR console

You can refer to the following steps for assigning the security groups when you create a new cluster using advanced options:

1. Navigate to the EMR console at `https://console.aws.amazon.com/elasticmapreduce/`.

2. Choose **Create cluster** and go to **Advanced options**.

3. Select the required options for your cluster until you reach **Step 4: Security section**.

4. Expand the **EC2 Security Groups** section on the page, which will show managed security groups and additional security groups for **Master** and **Core & Task** node types. By default, EMR managed security groups are selected and additional security groups are empty.

5. For EMR-managed security groups, if you would like to use your custom-managed security groups, then select them from the **EMR managed security groups** drop-down list.

   If you have selected a custom-managed security group, you will receive a message that requests you to choose a custom security group for other instances as you are not allowed to use a mix of managed and custom security groups for a cluster.

6. Optionally, under the **Additional security groups** section, select the pencil icon that will allow you to select up to a maximum of four security groups and then select **Assign security groups**. You need to repeat this for each **Master** and **Core & Task** node type.

7. After selecting other security configurations, select **Create Cluster**.

Now that you have understood how you can specify a security group using the EMR console, let's understand how you can do it with the AWS CLI command.

Specifying a security group using the AWS CLI

The following AWS CLI command provides an example that includes both EMR managed security groups and additional custom security groups for a cluster that will be launched in a private subnet of the VPC.

Refer to the `--ec2-attributes` parameter, which includes different security groups:

```
aws emr create-cluster --name "ClusterSecurityGroup" \
--release-label emr-emr-6.3.0 --applications Name=Hue Name=Hive
\
Name=Spark --use-default-roles --ec2-attributes \
SubnetIds=<subnet-xxxxxxxxxxxx>,KeyName=<myEC2KeyPair>,\
ServiceAccessSecurityGroup=<sg-xxxxxxxxxxxx>,\
EmrManagedMasterSecurityGroup=<sg-xxxxxxxxxxxx>,\
EmrManagedSlaveSecurityGroup=<sg-xxxxxxxxxxxx>,\
AdditionalMasterSecurityGroups=['<sg-xxxxxxxxxxxx>',\
'<sg-xxxxxxxxxxxx>','<sg-xxxxxxxxxxxx>'],\
AdditionalSlaveSecurityGroups=<sg-xxxxxxxxxxxx> \
--instance-type m5.2xlarge
```

Please replace the `<subnet-xxxxxxxxxxxx>`, `<myEC2KeyPair>`, and `<sg-xxxxxxxxxxxx>` variables before executing the command. Please note that \ is used to include newline characters in the command.

Specifying security groups for EMR notebooks

When you use or create an EMR notebook, EMR uses two security groups to control network traffic between your EMR notebook and cluster. The default security group that EMR uses has minimal rules to only allow communication between your notebook and the clusters to which it's attached.

EMR notebooks use Apache Livy to interact with the EMR cluster through a proxy that uses the `18888` TCP port. If you require access to be restricted to a subset of notebooks, then you can create your custom security group that has rules as regards imposing the restriction.

The following are the two security groups that EMR Notebooks uses to communicate with the EMR cluster:

- **ElasticMapReduceEditors-Livy**: In addition to the default security group attached to the master node of the cluster, this additional security group gets attached, which allows inbound traffic on TCP 18888 port.

- **ElasticMapReduceEditors-Editor**: This is the default security group attached to the EMR notebook, which allows outbound traffic from the notebook to the EMR cluster.

Apart from these two security groups, you might need to add an additional security group or rule if you plan to integrate the GitHub repository with your notebook. To access the GitHub repository from the notebook, you need to allow outbound traffic to the GitHub repository. It is recommended that you create a new custom security group for it and attach it to your notebook. If you plan to update the default `ElasticMapReduceEditors-Editor` security group, then all other notebooks that are attached to the default security group will also have the same access that you may not wish to give.

Connecting to Amazon EMR on an EC2 cluster using an interface VPC endpoint

You can connect to your VPC privately with other supported AWS services in the same or other AWS accounts, supported AWS Marketplace services, or VPC endpoint services with the help of AWS PrivateLink. PrivateLink is readily available and highly scalable.

With VPC endpoints, you can connect to your VPC-based EMR on an EC2 cluster using an AWS network and can avoid the internet route for connectivity, which provides better performance and security. You also don't need a NAT device, internet gateway, AWS Direct Connect or VPN connection for connectivity, and the EC2 instances in your VPC do not need any public IP addresses for connectivity, and the EC2 instances in your VPC do not need any public IP addresses to interact with the EMR API.

Each VPC endpoint represents one or more **elastic network interfaces** (**ENIs**) that have private IP addresses in VPC subnets. For connecting to your EMR cluster, you can use the AWS CLI or AWS console to create an interface VPC endpoint.

After you have created an interface VPC endpoint, one of the following happens:

- If you have enabled private DNS hostnames for your endpoint, then the default Amazon EMR endpoint resolves to your VPC endpoint. The default service name endpoint for EMR has the `elasticmapreduce.Region.amazonaws.com` format.

- If you have not enabled private DNS hostnames, then Amazon VPC provides a DNS endpoint name with the `VPC_Endpoint_ID.elasticmapreduce.Region.vpce.amazonaws.com` format.

For granular permission management for IAM users or groups, you can attach VPC endpoint-related IAM policies to your VPC endpoint. You can also attach security groups to your VPC endpoint to control inbound and outbound network traffic.

The following is an example of an IAM policy that allows VPC endpoint access to a specific IAM user in a particular AWS account, while access is denied to all remaining IAM users:

```
{
    "Statement": [
        {
            "Action": "*",
            "Effect": "Allow",
            "Resource": "*",
            "Principal": {
                "AWS": [
                    "arn:aws:iam::<AWS-Account-ID>:user/<IAM-
User-ID>"
                ]
            }
        }]
}
```

Please replace the `<AWS-Account-ID>` and `<IAM-User-ID>` variables with your AWS account and username, respectively.

Connecting to Amazon EMR on an EKS cluster using an interface VPC endpoint

Similar to EMR on the EC2 cluster, you can use VPC endpoints to connect to your EMR on an EKS cluster using the Amazon network and thereby avoid going through the public internet. If you create an EC2 instance in your VPC public subnet, then it can connect to EMR on the EKS API without the need for a public IP address.

As highlighted in the previous section, you can create an interface VPC endpoint using the EMR console or AWS CLI.

After you have created an interface VPC endpoint:

- If you have enabled private DNS hostnames for your endpoint, then the default Amazon EMR on the EKS endpoint resolves to your VPC endpoint. The default format for EMR on the EKS service name endpoint is `emr-containers.Region.amazonaws.com`.

- If you have not enabled private DNS hostnames, then Amazon VPC provides a DNS endpoint name with the `VPC_Endpoint_ID.emr-containers.Region.vpce.amazonaws.com` format.

The following is an example of a VPC endpoint IAM policy that allows a specific AWS account to perform read-only operations on your EMR on the EKS cluster:

```
{
    "Statement": [
        {
            "Action": [
                "emr-containers:DescribeJobRun",
                "emr-containers:DescribeVirtualCluster",
                "emr-containers:ListJobRuns",
                "emr-containers:ListTagsForResource",
                "emr-containers:ListVirtualClusters"
            ],
            "Effect": "Allow",
            "Resource": "*",
            "Principal": {
                "AWS": [
                    "<AWS-Account-ID>"
                ]
```

```
            }
         }
      ]
}
```

Please replace the `<AWS-Account-ID>` variable with your AWS account before integrating it with your user or role.

In this section, you have learned about how you can control network traffic to your cluster using security groups and how you can connect to your EMR cluster using interface VPC endpoints.

Summary

Over the course of this chapter, you got an overview of the basics of security, which included creating a security configuration and assigning it to multiple EMR clusters. Then you learned how you can enable authentication and authorization for EMR APIs using AWS IAM users, groups, policies, and roles.

Then we dived deep into data protection, which included encrypting your data at rest in a cluster's local disk, Amazon S3, and also securing your data while in transit during distributed processing.

Finally, we covered how you can configure managed and custom security groups for your cluster nodes and how configuring interface VPC endpoints can provide better security and performance.

That concludes this chapter! In the next chapter, we will dive deep into data-level security where you will learn how you can enable granular permission management on your cluster data using AWS Lake Formation and Apache Ranger.

Test your knowledge

Before moving on to the next chapter, test your knowledge with the following questions:

1. Assume that as part of your EMR cluster, you have some custom applications running that will be interacting with AWS services directly instead of executing Hadoop or Spark jobs. Your custom application needs to authenticate itself with AWS IAM to interact with the AWS services and should also have required privileges. How would you enable your application to authenticate itself with AWS IAM to get temporary credentials for access?

2. Assume that you are using Amazon S3 as your persistent data store in EMR and your organization has strict security rules to encrypt all the data you store. You have your own custom encryption keys that need to be used to encrypt your data. How would you ensure that EMR uses your custom key to encrypt data at rest?

3. Assume that you have an EMR notebook that needs to push or pull code from the GitHub repository and you have required IAM privileges to modify the default security groups for your notebook. How would you configure security group rules to allow access to your public GitHub repository?

Further reading

The following are a few resources you can refer to for further reading:

- Learn how to configure a Kerberos configuration in EMR: `https://docs.aws.amazon.com/emr/latest/ManagementGuide/emr-kerberos.html`

- Creating keys and certificates for encryption: `https://docs.aws.amazon.com/emr/latest/ManagementGuide/emr-encryption-enable.html`

- Setting up cross-account access for EMR on an EKS cluster: `https://docs.aws.amazon.com/emr/latest/EMR-on-EKS-DevelopmentGuide/security-cross-account.html`

8

Understanding Data Governance in Amazon EMR

In previous chapters, you learned about **EMR cluster** security with IAM policies and data encryption and how you can configure security groups to control network traffic from or to your cluster.

As well as EMR **cluster-level security**, you can also enable **data-level security** where you can build a centralized data catalog on your datasets and then define fine-grained permissions to control which user can access which database, table, or column of your data catalog. Security of data is as important as maintaining security on your infrastructure. When you put security controls on your data, you also need to think about whether the data available for consumption is available in a useful format with proper data quality checks in place.

That brings us to the focus of this chapter, where we will dive deep into the following topics, which will help you implement data governance and granular permission management on your data catalog:

- Understanding data catalog and access management options
- Understanding Amazon EMR integration with AWS Lake Formation
- Understanding Amazon EMR integration with Apache Ranger

This will help your organization to build a data governance strategy, where they can put controls around the data catalog and security around its access.

Technical requirements

In this chapter, we will dive deep into the EMR cluster's integration with AWS Glue Data Catalog and the AWS Lake Formation service. So, to test the integration, you will need the following resources before you get started:

- An AWS account
- An **Identity and Access Management** (**IAM**) user that has permission to create and manage an EMR cluster with related resources, such as Amazon **Elastic Compute Cloud** (**EC2**) instances, required IAM roles, and security groups
- IAM access privileges to integrate AWS Glue Data Catalog, AWS Lake Formation, Amazon **Simple Storage Service** (**S3**), CloudWatch, and CloudTrail

Now let's understand how you can build a centralized data catalog in EMR and what options you have for this integration.

Understanding data catalog and access management options

When you think of **data lake** use cases, where the storage layer is a filesystem such as HDFS or an object store such as Amazon S3, by default, the data is not represented as databases or tables. In a data lake, you may receive datasets as structured, semi-structured, or unstructured datasets or files.

If it is *unstructured* data, such as media files (images, videos), then often machine learning or artificial intelligence tools are integrated to extract data and metadata about the media files and save the output to a data lake for further analytics.

If it is *semi-structured*, then often it goes through **Extract, Transform, and Load** (ETL) transformations to flatten it so that it is available to data analysts or data scientists for consumption.

Structured data, which is available as files or objects in a data lake, is not accessible to business users or data analysts in a form that they can query data using standard SQL. To make the data available as databases or tables for business users, you can think of creating a virtual table that imposes the schema while reading the datasets.

In an ideal database world, data that gets written to databases would follow a **schema-on-write** approach whereas in a data lake it's primarily **schema-on-read**, which means when you submit a query to read the data, the schema is applied on top of the filesystem to show the output in a tabular format. Whereas for schema-on-write semantics, before the data gets written to the database storage, its schema is validated against the table schema, and upon validation, it gets written to the storage.

When you integrate Amazon EMR for your data analytics use cases, you can store the data in either an EMR cluster's HDFS or Amazon S3 using EMRFS. Amazon S3 is the recommended storage as it provides high availability and scalability. On top of the data store, if you need to create virtual tables, then you have the following options:

- **Hive Metastore**: You can integrate Apache Hive while creating your EMR cluster, which uses a relational database running on the cluster master node as its metastore, or you can integrate Amazon RDS as its external metastore for better reliability or cross-cluster metadata sharing. Then, you can create databases and virtual tables on top of the cluster HDFS path or Amazon S3 path and run standard SQL queries to fetch data.

- **AWS Glue Data Catalog**: This is a serverless managed service that is designed to act as the centralized data catalog on top of the S3 data lake as well as other AWS services, such as Amazon Redshift, Amazon DynamoDB, and relational databases connected through JDBC.

If you have configured Amazon S3 as your cluster's persistent data store, then AWS Glue Data Catalog is the recommended option as that provides the opportunity for additional integrations. As an example, you can integrate AWS Glue ETL jobs on top of an S3 data lake using Glue Data Catalog tables, integrate AWS Lake Formation granular permission management, or enable cross-account data sharing for centralized data management.

Now, let's dive deep into AWS Glue Catalog and understand how you can integrate that with your EMR cluster as an external metastore.

Using AWS Glue Data Catalog

AWS Glue Data Catalog is a persistent metastore that allows you to build a centralized data catalog that can be shared across multiple AWS analytics services and can also be shared between multiple AWS accounts. It is integrated with AWS IAM, using which you can control which user is allowed to invoke Glue Data Catalog APIs, such as creating databases or creating tables.

In a data lake use case, AWS Glue crawlers play an important role of crawling subset data from a specified Amazon S3 path to autodetect the schema and create metadata tables in Glue Data Catalog. Glue Data Catalog also has audit and data governance capabilities that keep track of schema changes and create a new version with each update.

The following are the AWS services that are integrated with AWS Glue Data Catalog:

- **AWS Glue jobs**: AWS Glue ETL jobs read from the catalog for ETL processing and also update the output to Glue Data Catalog.

- **Amazon Athena**: Glue Data Catalog is one of the primary data sources for Athena, where Glue Data Catalog databases and tables are listed for you to query using standard SQL.

- **Amazon Redshift Spectrum**: Similar to Amazon Athena, Redshift Spectrum can fetch data from a data lake and other sources by querying through Glue Data Catalog databases and tables.

- **Amazon EMR**: As described earlier, similar to Hive Metastore, EMR can use Glue Data Catalog as its external metastore.

- **AWS Lake Formation**: Using AWS Lake Formation, you can define granular permission management on top of your Glue Data Catalog databases, tables, columns, or rows. Once you enable Lake Formation on your AWS account and integrate permission management, other AWS services, such as Athena, Glue jobs, Redshift Spectrum, and EMR, follow the access policies defined in Lake Formation.

After understanding what the role of Glue Data Catalog is, let's learn how you can integrate Glue Data Catalog in Amazon EMR.

Integrating AWS Glue Data Catalog with Amazon EMR

As explained in *Chapter 5*, *Setting Up and Configuring Clusters*, when you create your EMR cluster using advanced options, on the **Step 1: Software and Steps** screen, you have optional **AWS Glue Data Catalog** settings, which allow you to configure Glue Data Catalog for Hive, Presto, and Spark SQL.

The following screenshot shows the settings in the EMR console:

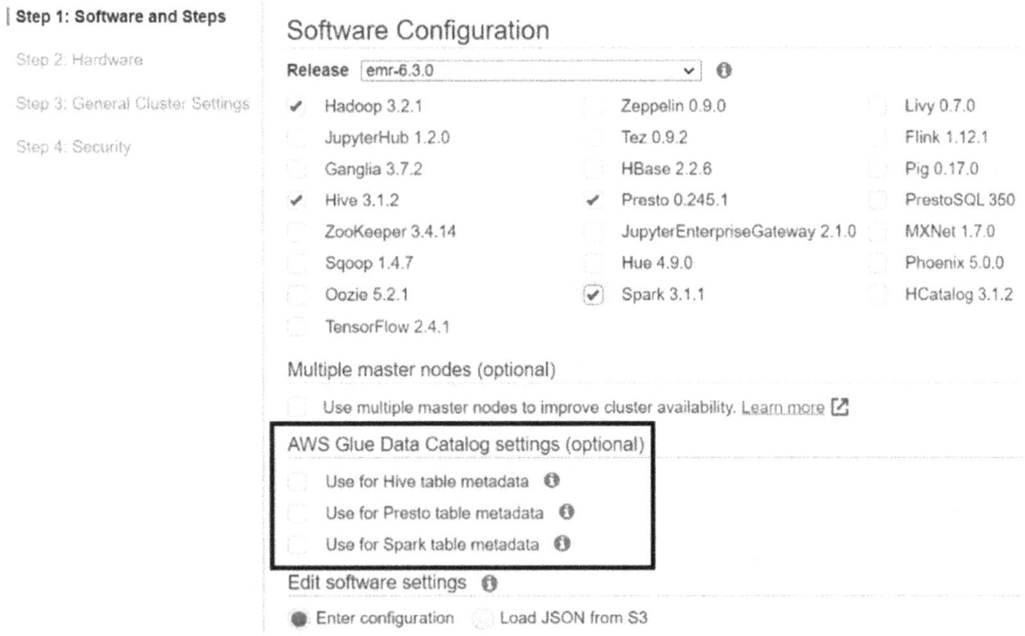

Figure 8.1 – EMR console to configure Glue Data Catalog

You can enable the same settings with the AWS **Command Line Interface** (**CLI**) and the following is an example of it:

```
aws emr create-cluster --name 'EMR with Glue Catalog'
--applications Name=Hadoop Name=Hive Name=Presto
Name=Spark --release-label emr-6.3.0 --configurations
'[{"Classification":"hive-site","Properties":{"hive.
Metastore.client.factory.class":"com.amazonaws.glue.catalog.
Metastore.AWSGlueDataCatalogHiveClientFactory"}},{"Class
ification":"presto-connector-hive","Properties":{"hive.
Metastore.glue.datacatalog.enabled":"true"}},{"Classific
ation":"spark-hive-site","Properties":{"hive.Metastore.
client.factory.class":"com.amazonaws.glue.catalog.Metastore.
AWSGlueDataCatalogHiveClientFactory"}}]' --use-default-roles
--region us-east-1
```

As you can see, the --configurations parameter in this command has the configurations that specify Glue Data Catalog for Hive, Spark, and Presto. We have explained the Glue Data Catalog integration with Hive, Presto, and Spark SQL in detail in *Chapter 4*, *Big Data Applications and Notebooks Available in Amazon EMR*.

Permission management on top of a data catalog

After customers start using EMR with HDFS or EMRFS with S3 as their distributed storage layer for big data processing, the next thing they look for is data governance and granular permission management on their data lake. This will enable them to provide database-, column-, or row-level permissions on top of Hive Metastore or Glue Catalog.

To implement permission management in EMR, you have the following options:

- AWS Lake Formation
- Apache Ranger

Now, let's dive into each of these options and understand how you can integrate them with EMR.

Understanding Amazon EMR integration with AWS Lake Formation

AWS Lake Formation is a managed service using which you can control which user can access which databases, tables, columns, or rows of your table. Lake Formation also supports integration with **Active Directory Federation Services** (**AD FS**) and SAML-based **single sign-on** (**SSO**), which allows users to authenticate themselves using their organization's login credentials.

AWS Lake Formation has several features; the following are a few of the popular features:

- **Blueprints**: Lake Formation blueprints provide you with a few templates using which you can ingest data from relational databases or AWS load balancer logs to an S3 data lake. It invokes AWS Glue workflows and jobs to do the data ingestion.

- **Granular permission management**: With Lake Formation permissions management, you can define database, table, column, or row-level permissions that validate every user request. You can also define Lake Formation tags and define tag-based permissions, instead of defining access for every database or table.

- **Lake Formation governed tables**: With Lake Formation governed tables, you can do row-level transaction updates on your Glue Data Catalog tables, which will enable updates or merges on your S3 data lake objects. This is great for GDPR compliance requirements, which require data to be updated or deleted. Lake Formation governed tables also have features such as query acceleration with predicate pushdown, storage optimization with auto compression of small S3 files, and time travel, using which you can access snapshots of your data at a specific time in the past.

Out of all the preceding features, we will primarily focus on Lake Formation fine-grained permission management and understand how you can integrate it with Amazon EMR.

We assume Lake Formation is enabled on your account and you have defined granular permissions on your Glue Data Catalog tables. Now, when you run queries on top of these Glue Data Catalog tables using any of the AWS analytics services, such as Amazon Athena, Amazon Redshift, AWS Glue jobs, Amazon QuickSight, or Amazon EMR, Lake Formation permissions come into play to allow or deny the request.

The following diagram explains how Lake Formation works when a user submits a query using these AWS services:

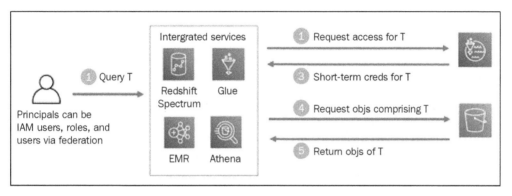

Figure 8.2 – AWS Lake Formation – user request processing

As you can see in this diagram, the user submits a query to Amazon EMR, Redshift Spectrum, AWS Glue, or Amazon Athena to fetch data from the data lake. AWS Lake Formation validates this request and if allowed, it generates a short-term credential that the AWS analytics service can use to retrieve data from the data lake and return it to the user.

Now let's understand how Lake Formation integration works with Amazon EMR.

Integrating Lake Formation with Amazon EMR

Starting from EMR release 5.31.0, you can launch a cluster with AWS Lake Formation integration, which provides the following two key benefits:

- Granular permission on Glue Data Catalog databases and tables
- SAML-based federated SSO to your EMR notebooks or Apache Zeppelin notebook using your corporate credentials

Now let's understand how you can launch an EMR cluster with Lake Formation.

IAM role needed for Lake Formation setup

The following are the three key IAM roles you need to set up for EMR to work with Lake Formation:

- **Custom EC2 instance profile role**: To make EMR work with Lake Formation, please make sure you create a custom EC2 instance profile so that you can edit or add policies for Lake Formation integration.

- **Additional IAM role for Lake Formation**: This IAM role for Lake Formation defines which **identity providers** (**IdPs**) can assume this role and what privileges a user will have when they log in through an **IdP**.

- **IAM role for non-Lake Formation AWS services**: This role will be used by EMR to interact with AWS services that are not integrated with Lake Formation, such as DynamoDB and Kinesis Data Streams. This role should not include any AWS Glue or Lake Formation API operations, any AWS **Security Token Service** (**STS**) `AssumeRole` operations, or any Amazon S3 bucket or prefix that is controlled by AWS Lake Formation. For S3 paths registered with Lake Formation, EMR will use the IAM role that is integrated with Lake Formation.

We suggest you read through the AWS documentation (link in the *Further reading* section) to understand the Lake Formation setup steps you will need to configure before you begin integrating Lake Formation as part of your EMR cluster.

Next, let's learn about a few EMR components that help with Lake Formation fine-grained access control.

EMR components that help with Lake Formation integration

Amazon EMR uses the following key components to facilitate integration with Lake Formation:

- **Proxy agent**: This is an Apache `knox`-based agent that runs on the EMR master node as the `knox` system user. This agent helps generate temporary credentials when it receives SAML-authenticated requests from users. While running, it writes logs to the `/var/log/knox` directory of the master node.

- **Secret agent**: This agent runs on every node of the cluster and uses Glue APIs to retrieve Glue Data Catalog metadata information and Lake Formation APIs to get temporary credentials to provide access. This agent securely stores secrets such as user temporary credentials or encryption keys of Kerberos tickets and distributes them to other EMR applications for authentication or authorization. This runs as the `emrsecretagent` user on cluster nodes and writes its logs to the `/emr/secretagent/log` directory

Please note, this agent process is dependent on a set of `iptable` rules, so make sure that `iptable` is not disabled and you have not altered the rules if you customized it.

- **Record server**: Similar to the secret agent, this process also runs in every node of the cluster and is named as the `emr_record_server` user. It uses the temporary credentials distributed by the secret agent to authorize requests and then reads data from the S3 data lake as per the row or column-level access defined in Lake Formation. This writes logs to the `/var/log/emr-record-server` directory of the nodes.

The following is an architecture reference diagram that explains how these three components work to provide SSO capability with SAML authentication and how Lake Formation is integrated to provide fine-grained access control with Amazon S3:

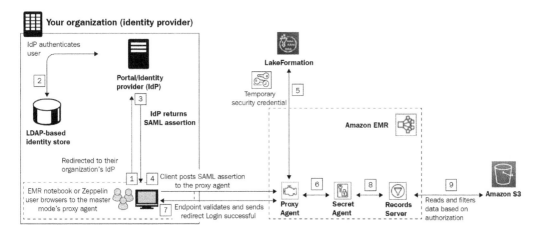

Figure 8.3 – Architecture reference for SAML-based authentication in EMR

From a user standpoint, the SAML-based authentication and Lake Formation based authorization work seamlessly such that users need not provide their credentials and it automatically signs in when they are accessing EMR notebooks or Zeppelin notebooks.

After getting an overview of the Lake Formation way of working with EMR, now let's understand how you can launch an EMR cluster with Lake Formation.

Launching an EMR cluster with Lake Formation

Please make sure you have followed the setup steps and prerequisites specified in the AWS documentation (`https://docs.aws.amazon.com/emr/latest/ManagementGuide/emr-lf-prerequisites.html`).

Apart from creating a custom EC2 instance profile role, please make sure you have created a security configuration that enables Lake Formation configuration.

The following screenshot shows how you can enable it using the EMR console:

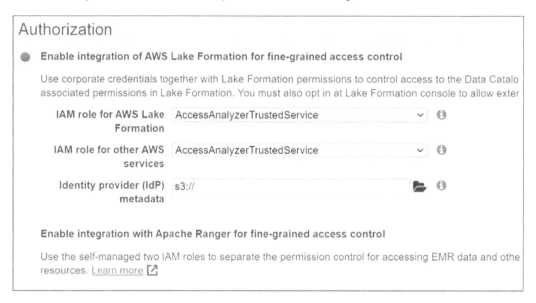

Figure 8.4 – EMR security configurations with Lake Formation

After your security configuration is ready, you can launch an EMR cluster using the following AWS CLI command, which includes a custom EC2 instance profile role, the security configuration name that you created, and the --kerberos-attributes parameter if your cluster has Kerberos configuration enabled.

This cluster enables Zeppelin integration with Lake Formation:

```
aws emr create-cluster --region us-east-1
--name emr-lakeformation --release-label
emr-6.3.0 --use-default-roles --instance-groups
InstanceGroupType=MASTER,InstanceCount=1,InstanceType=m4.
2xlarge
InstanceGroupType=CORE,InstanceCount=1,InstanceType=m4.2xlarge
--applications Name=Zeppelin Name=Livy --kerberos-attributes
Realm=EC2.INTERNAL,KdcAdminPassword=<MyClusterKDCAdminPassword>
--ec2-attributes KeyName=<MyEC2KeyPair>,SubnetId=<subnet-00xxxx
xxxxxxxxx11>,InstanceProfile=<MyCustomEC2InstanceProfile>
--security-configuration <security-configuration-name>
```

Please replace the `<MyClusterKDCAdminPassword>`, `<MyEC2KeyPair>`, `<subnet-00xxxxxxxxxxxxxx11>`, `<MyCustomEC2InstanceProfile>`, and `<security-configuration-name>` variables before executing the command.

If you have configured Active Directory authentication with SSO, then as a next step, you should update the SSO URL for your IdP, as we will see in the following section.

Updating the SSO URL with your IdP

Please refer to the following steps to update the callback or SSO URL so that your users can be redirected to the EMR cluster's master node DNS URL:

If you are using **Active Directory Federation Services** (**AD FS**) as your IdP, then do the following:

1. From the AD FS management console, navigate to **Relying Party Trusts**.
2. Right-click on the display name of your replying party trust and select **Properties**.
3. From the **Properties** window, select the **Endpoints** tab.
4. Select **Edit** for the temporary URL you provided earlier.
5. In the **Edit endpoint** window, replace the trusted URL with your EMR cluster's master node DNS.
6. In the **Add an endpoint** window, enter your EMR cluster's master node public DNS in the **Trusted URL** field, which might look like `https://ec2-11-111-11-111.compute-1.amazonaws.com:8442/gateway/knoxsso/api/v1/websso?pac4jCallback=true&client_name=SAML2Client`.
7. Then, click **OK**.

This is just an example for AD FS. For any other IdPs, such as Okta or Azure Active Directory, you can follow the steps given by the respective IdP.

Setting up EMR notebooks to work with Lake Formation

After your cluster is launched with Lake Formation integration, you can use an EMR notebook or Zeppelin for interactive development. Before accessing these notebook interfaces, make sure your cluster's **network access control list** (**NACL**) and cluster security group have allowed access to port `8442` from your local system IP.

> **Important Note**
>
> By default, the EMR cluster's proxy agent uses a self-signed TLS certificate, so while accessing the notebook URLs, your browser will have the warning to accept the certificate to continue accessing the URL. But you can apply a custom certificate to your proxy agent.

Now let's understand how you can access both of these notebooks.

Accessing Apache Zeppelin

After your cluster is launched, you can get the cluster's master node public DNS from the EMR console. Then, you can access Zeppelin by using the `https://<MasterNodePublicDNS>:8442/gateway/default/zeppelin/` URL.

As described, your browser will prompt you to accept the self-signed certificate. If you have integrated IdP, then after you accept the certificate, it will redirect you to your IdP, where you can authenticate yourself and then get automatically redirected to Zeppelin.

In the Zeppelin interface, you can create a new notebook and then use Spark SQL to access Lake Formation databases or tables.

Accessing EMR notebooks

You can create an EMR notebook using the EMR console and integrate the notebook with an existing EMR cluster that has enabled Lake Formation.

In the EMR console, you can navigate to **Notebooks | Create Notebook** and then attach the notebook to an EMR cluster. Similar to Zeppelin, after accepting the self-signed certificate, you will be redirected to your IdP. Once authenticated, it will automatically redirect to your EMR notebook.

This concludes the Lake Formation integration with Amazon EMR. Next, we can see how Apache Ranger is integrated with EMR to provide fine-grained access control.

Understanding Amazon EMR integration with Apache Ranger

Apache Ranger is an open source framework that provides comprehensive security across the Hadoop ecosystem, using which you can define and manage security policies to control access on Hadoop components.

Starting from the EMR 5.32.0 release, your EMR cluster has default native integration with Apache Ranger. That means EMR installs and manages the Ranger plugin on your behalf.

Similar to AWS Lake Formation, Apache Ranger also provides fine-grained access control on top of Hive Metastore or Amazon S3 prefixes. Using Ranger, you can define access permissions on top of Hive databases, tables, or columns while using Hive queries or Spark jobs. Data masking and row-level filtering are only supported with Hive.

Ranger has the following two primary components:

- **Apache Ranger policy admin server**: With this server, you can define authorization policies for Hive Metastore, Apache Spark, and EMRFS with S3. To integrate with EMR, you can use your existing Ranger policy admin server or set up a new one.

- **Apache Ranger plugin**: This component helps in validating user access against the policies defined in the Ranger policy admin server.

The following diagram explains the Apache Ranger architecture diagram in EMR:

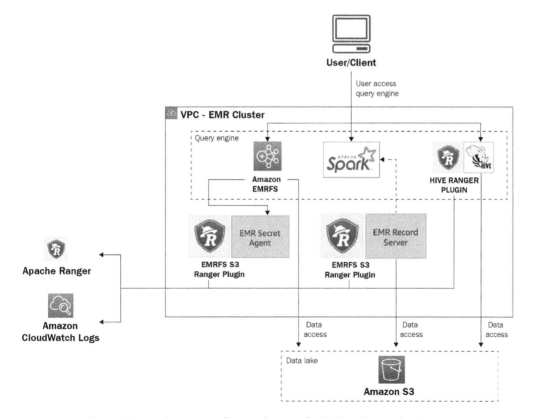

Figure 8.5 – Architecture reference diagram for EMR with Apache Ranger

As you can see in this architecture diagram, EMR uses the following two components to work with Apache Ranger:

- **EMR secret agent**: As explained when discussing AWS Lake Formation integration with EMR, this agent stores and distributes secrets, including user credentials, Kerberos tickets, or encryption keys. The secret agent validates user requests and generates temporary credentials for access.

- **EMR record server**: As explained in the *Integrating Lake Formation with Amazon EMR* section, this runs in every node of the cluster and uses the temporary credentials to authorize the request, and then retrieves authorized data from S3.

By default, Amazon EMR supports Ranger integration with Spark, Hive, and EMRFS + S3. Starting from the EMR 5.32.0 release, you can enable Ranger for other EMR components, such as Apache Hadoop, Apache Livy, Apache Zeppelin, Apache Hue, Tez, Ganglia, ZooKeeper, MXNet, Mahout, HCatalog, and TensorFlow with additional configuration.

Now let's learn how you can set up Ranger in an EMR cluster.

Setting up Apache Ranger in EMR

To set up Apache Ranger in EMR, the following are some of the steps you should consider.

Setting up the Ranger admin server

The Apache Ranger plugin in EMR uses SSL/TLS to interact with the admin server. To enable SSL/TLS, you need to configure the following attribute in the `ranger-admin-site.xml` file on the admin server:

```
<property>
    <name>ranger.service.https.attrib.ssl.enabled</name>
    <value>true</value>
</property>
```

Apart from the preceding SSL configuration, you also need to configure the following additional configurations:

```
<property>
    <name>ranger.https.attrib.keystore.file</name>
    <value>_<PATH_TO_KEYSTORE>_</value>
</property>
<property>
```

```
    <name>ranger.service.https.attrib.keystore.file</name>
    <value>_<PATH_TO_KEYSTORE>_</value>
</property>
<property>
    <name>ranger.service.https.attrib.keystore.pass</name>
    <value>_<KEYSTORE_PASSWORD>_</value>
</property>
<property>
    <name>ranger.service.https.attrib.keystore.keyalias</name>
    <value><PRIVATE_CERTIFICATE_KEY_ALIAS></value>
</property>
<property>
    <name>ranger.service.https.attrib.clientAuth</name>
    <value>want</value>
</property>
<property>
    <name>ranger.service.https.port</name>
    <value>6182</value>
</property>
```

With these configuration parameters, you can provide details about your certificate, including the certificate alias, path, password, and ranger service port.

IAM roles for native integration to set up the Ranger admin server

Before launching your cluster, you need to create the following roles that Apache Ranger uses:

- **Custom EC2 instance profile role**: Instead of using the default EMR_EC2_ DefaultRole role that we explained in *Chapter 5, Setting Up and Configuring Clusters*, and *Chapter 7, Understanding Security in Amazon EMR*, you need to create a custom role that should have permission to tag sessions and access TLS certificates available in AWS Secrets Manager.

- **IAM role for Apache Ranger**: This role provides temporary credentials using which the EMR record server and Hive can access S3 data. Please make sure to include access to **Key Management Service** (**KMS**) keys, if you have enabled encryption on your S3 bucket using S3-SSE. In addition, you also need to create a trust policy between the EC2 instance profile and this role so that your instance can assume this role. You can refer to the AWS IAM documentation to learn how you can configure the trust policy.

- **IAM role for other services**: If needed, this role is used to interact with other AWS services. Similar to the IAM role for Apache Ranger, please add a trust policy between this role and the EC2 instance profile so that the EC2 instance can assume this role to interact with other AWS services, such as Amazon Kinesis Data Streams and Amazon DynamoDB.

For a complete list of IAM policies that will be embedded into any of the preceding roles, please refer to the AWS documentation.

Storing TLS certificates in AWS Secrets Manager

As explained in the previous section, the Ranger admin server communicates with EMR over TLS to make sure the communication is secure and cannot be intercepted if read by unauthorized processes. It is mandatory that Ranger plugins for Hive, Spark, or S3 authenticate to EMR using two-way TLS authentication, which requires two public and two private certificates. You must use AWS Secrets Manager to configure these TLS certificates and then integrate them into EMR security configurations.

> **Important Note**
> It is recommended that you generate a separate set of TLS certificates for each Ranger plugin so that if one of the plugin keys is compromised, you are not risking all plugins.
>
> Also, you should rotate your certificates before expiry to continue having access.

EMR security configurations for Apache Ranger

After you have created the required roles and trust policies, you can create EMR security configurations that enable Apache Ranger fine-grained access control.

The following screenshot shows how you can enable it using the EMR console:

○ **Enable integration with Apache Ranger for fine-grained access control**

Use the self-managed two IAM roles to separate the permission control for accessing EMR data and other AWS services. resources. Learn more [↗]

AWS IAM configurations

EMR recommends two IAM roles to separate the permission control for accessing EMR data and other AWS services. Spec Learn more [↗]

IAM role for Apache Ranger   Provide an IAM role for access governed by Apache Ranger access control policies.

| AccessAnalyzerTrustedService | ∨ |

IAM role for other AWS Services   Provide an IAM role for access governed by IAM permission only.

| AccessAnalyzerTrustedService | ∨ |

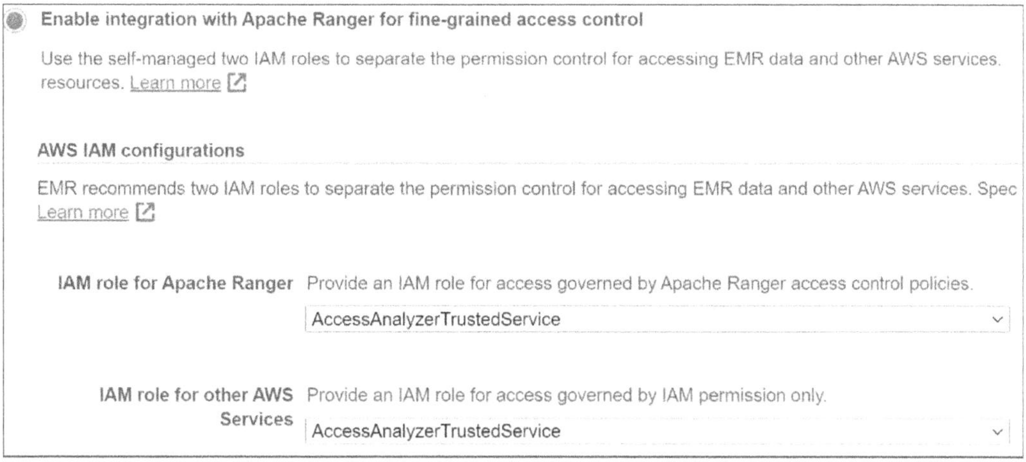

Figure 8.6 – EMR security configurations to enable Apache Ranger

After you have created the security configuration, you can attach it to your EMR cluster while launching it using the EMR console or AWS CLI.

Understanding Apache Ranger plugins

Starting from the EMR 5.32 release, EMR includes the following Ranger plugins, which integrate with Ranger 2.0 to provide fine-grained access control and audit capabilities. These plugins validate access against the policies defined in the Ranger policy admin server.

Now, let's get an overview of each of these plugins.

Ranger plugin for Hive

In EMR, the Ranger plugin for Hive supports all the functionality available in the open source version, which includes database-, table-, column-, and row-level permissions with the data masking feature.

The Hive plugin is, by default, compatible and integrated with the existing Hive service definition. In the Ranger console, if you do not find an instance of the Hive service under **Hadoop SQL**, then please click the + icon next to it and add the service name as amazonemrhive. You will need this service name while creating the EMR security configurations.

Additionally, you need to configure connection properties for the Ranger admin server to connect with **HiveServer2**, and the properties include **Username**, **Password**, **jdbc.driverClassName**, **jdbc.url**, and **Common Name for Certificate**.

The following is a screenshot of the Ranger Service Manager console that shows the `amazonemrhive` configuration under **HADOOP SQL**:

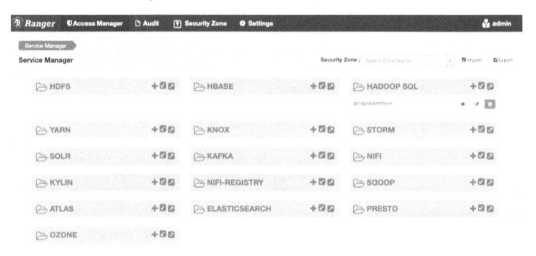

Figure 8.7 – Ranger console that shows amazonemrhive under Hadoop SQL

In this section, you have learned about the Ranger plugin for Hive and how you configure it. Next, you will learn how you can configure the Ranger plugin for the Spark engine.

Ranger plugin for Spark

In EMR, the Ranger plugin for Spark supports fine-grained access control on Spark SQL queries that query data from Hive Metastore. You can define access control on databases, tables, or the column level.

When a Spark executor runs a SparkSQL query, it goes through the record server to validate access defined in the Ranger policy admin server. In your Ranger policies, you can include grant or deny policies for users or groups and also log audit events to Amazon CloudWatch.

Please refer to the AWS documentation for the complete setup steps.

Ranger plugin for EMRFS S3

EMR uses EMRFS to interact with Amazon S3. When you try to access data from S3, it goes through the following steps:

1. EMRFS sends a request to the secret agent to get temporary credentials.

2. This request gets authorized against the Ranger plugin.

3. If the request is authorized, then the secret agent assumes the IAM role for Ranger that has restricted access to generate temporary credentials. These credentials will have access only to the resources defined in the Ranger policy for which the access was authorized.

4. Finally, these credentials are passed back to EMRFS to access data from S3.

You can create policies that allow or deny access to specific users or groups and the policy can point to a specific S3 bucket or prefix.

For complete setup steps, refer to the AWS documentation.

This section provided an overview of Apache Ranger integration with EMR that included setting it up in EMR and understanding the Ranger plugin.

Summary

Over the course of this chapter, you got an overview of integrating a centralized data catalog on top of your distributed persistent storage layer using AWS Glue Data Catalog or Hive Metastore.

Then, you learned about how you can integrate fine-grained access control using AWS Lake Formation and Apache Ranger. This chapter provided an overview of the integration, its different components, and what some of the steps you should be taking to configure it are. The links provided in the *Further reading* section will guide you through the detailed configuration steps.

That concludes this chapter! Hopefully, this gives you a good starting point to integrate a centralized data catalog and data governance on top of your distributed data lake. In the next chapter, we will explain how you can implement a batch ETL use case using EMR.

Test your knowledge

Before moving on to the next chapter, test your knowledge with the following questions:

1. Assume you have multiple batch and streaming ETL workloads that use different transient EMR clusters for distributed processing. Your organization is looking for a persistent centralized data catalog that can help the data governance team get a unified view. Between AWS Glue Data Catalog and Hive Metastore, which one is better suited?

2. Assume you have an on-premises Hadoop cluster that uses Apache Ranger for fine-grained access control. You are planning to migrate your on-premises Hadoop cluster to Amazon EMR in AWS to take benefit of cloud security, reliability, and scaling capabilities. For your Ranger server, you have configured custom TLS certificates that you plan to integrate into EMR. How should you integrate the TLS certificates into EMR?

3. Assume you are part of a bigger enterprise that has multiple departments and each department has its own AWS account that owns its data. Your organization is looking for options using which they can build a centralized data catalog and permission management system that will be controlled by the data governance team. The central data governance team should be able to define permissions on all the data available in various AWS accounts and also be able to share catalog tables between accounts. Which architecture should you follow for centralized permission management and cross-account data sharing?

Further reading

The following are a few resources you can refer to for further reading:

*   Considerations and limitations for AWS Glue Data Catalog with Amazon EMR: `https://docs.aws.amazon.com/emr/latest/ReleaseGuide/emr-hive-Metastore-glue.html`

*   Setting up AWS Lake Formation: `https://docs.aws.amazon.com/lake-formation/latest/dg/getting-started-setup.html`

*   Detailed steps for Lake Formation integration with Amazon EMR: `https://docs.aws.amazon.com/emr/latest/ManagementGuide/emr-lake-formation.html`

*   Detail steps to configure Apache Ranger in EMR: `https://docs.aws.amazon.com/emr/latest/ManagementGuide/emr-ranger.html`

Section 3: Implementing Common Use Cases and Best Practices

This part of the book will explain how to implement the most common use cases of Amazon EMR, including batch ETL with Spark, real-time streaming with Spark Streaming, and handling UPSERT operations in S3 data lakes with Apache Hudi. Then it will explain how you can orchestrate your EMR jobs and how you can strategize on-premises Hadoop cluster migration to EMR, and finally, it will cover some of the best practices and cost optimization techniques you can follow while implementing your data analytics pipeline in EMR.

This section comprises the following chapters:

- *Chapter 9, Implementing Batch ETL Pipeline with Amazon EMR and Apache Spark*
- *Chapter 10, Implementing Real-Time Streaming with Amazon EMR and Spark Streaming*
- *Chapter 11, Implementing UPSERT on S3 Data Lake with Apache Spark and Apache Hudi*
- *Chapter 12, Orchestrating Amazon EMR Jobs with AWS Step Functions and Apache Airflow/MWAA*
- *Chapter 13, Migrating On-Premises Hadoop Workloads to Amazon EMR*
- *Chapter 14, Best Practices and Cost Optimization Techniques*

9

Implementing Batch ETL Pipeline with Amazon EMR and Apache Spark

In *Chapter 2, Exploring the Architecture and Deployment Options*, you learned about different EMR use cases such as batch **Extract, Transform, and Load** (ETL), real-time streaming with EMR and Spark streaming, data preparation for **machine learning** (ML) models, interactive analytics, and more.

In this chapter, we will dive deep into a use case – **Batch ETL with Amazon EMR and Apache Spark**, where we will look at the implementation steps that you can follow to replicate the setup in your AWS account.

We will cover the following topics, which will help you understand the use case, its application architecture, and how a transient EMR cluster with Spark can be integrated for distributed processing:

- Use case and architecture overview

- Implementation steps

- Validating output through Athena

- Spark ETL and Lambda function code walk-through

Batch ETL is a common use case across many organizations and this use case implementation learning will provide you with a starting point, using which you can build more complex data pipelines in AWS using Amazon EMR.

Technical requirements

In this chapter, we will implement a batch ETL pipeline using AWS services, so before getting started, make sure you have the following requirements:

- An AWS account with access to create Amazon S3, AWS Lambda, Amazon EMR, Amazon Athena, and AWS Glue Data Catalog resources

- An IAM user that has access to create IAM roles, which will be used to trigger or execute jobs

- Access to the GitHub repository:

  `https://github.com/PacktPublishing/Simplify-Big-Data-Analytics-with-Amazon-EMR-/tree/main/chapter_09`

Now let's dive deep into the use case and hands-on implementation steps.

Check out the following video to see the Code in Action at `https://bit.ly/3LtLZGX`

Use case and architecture overview

For this use case, let's assume you have a vendor who provides incremental sales data at the end of every day. The file arrives in S3 as CSV and it needs to be processed and made available to your data analysts for querying.

Your assignment is to build a data pipeline that automatically picks up the new sales file from the S3 input bucket, processes it with required transformations, and makes it available in the target S3 bucket, which will be used for querying. To implement this pipeline, you have planned to integrate a transient EMR cluster with Spark as the distributed processing engine. This EMR cluster is not active and gets created just before executing the job and gets terminated after completing the job.

Architecture overview

The following is the high-level architecture diagram of the data pipeline:

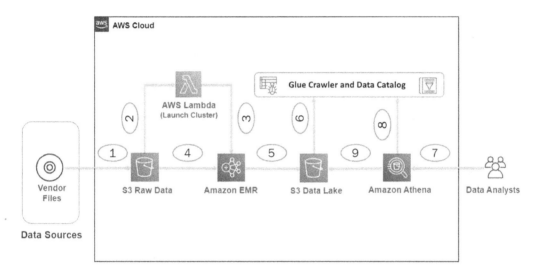

Figure 9.1 – Reference architecture diagram for a batch ETL pipeline

Here are the steps as shown in the previous diagram:

- *Step #1* represents the vendor pushing the sales CSV file to an Amazon S3 raw bucket.

- *Step #2* represents the triggering of an AWS Lambda function based on the CSV file's S3 PUT event.

- *Step #3* represents that the AWS Lambda function invokes the Amazon EMR API to launch a cluster with a Spark step.

- *Steps #4 and #5* represent the EMR cluster with the Spark job, which reads input sales data from the raw S3 bucket, does processing, and then writes the output to the S3 data lake bucket. The PySpark job renames the column names by replacing empty spaces with an underscore and writes output in Parquet format with the `/<year>/<month>/` partition columns.

- *Step #6* represents Glue Data Catalog tables defined on top of the data lake bucket, which will be used by Amazon Athena for querying.

- *Steps #7, #8, and #9* represent data analysts using Amazon Athena for querying data using standard SQL. Amazon Athena queries table metadata from Glue Data Catalog and then queries data from Amazon S3 to show the output in structured tabular format to the end user.

Having gotten an overview of the use case and architecture, let's get started on the implementation steps. Please make sure you meet all the prerequisites defined in the *Technical requirements* section of this chapter.

Implementation steps

In this section, we will guide you through the implementation steps for the use case and architecture we explained in the previous section.

> **Important Note**
> Please note, while explaining the implementation steps, we have used **us-east-1** as the AWS region. You can use the same or an alternate region as per your choice. Please check any resource or service limits that might apply to your AWS region before proceeding with the implementation.

Creating Amazon S3 buckets

Let's first create the Amazon S3 buckets and folders that will be used for both input and output. Please refer to the following steps to create them:

1. Navigate to the Amazon S3 console at `https://s3.console.aws.amazon.com/s3/home?region=us-east-1#`.

2. From the buckets list, choose **Create Bucket**, which will open up a form on the web interface to provide your bucket name and related configurations.

    We have specified the input bucket name as `raw-input` and kept everything else as the default.

3. Then click the **Create bucket** button to create the bucket.

The following screenshot shows the AWS console, using which we have created the bucket:

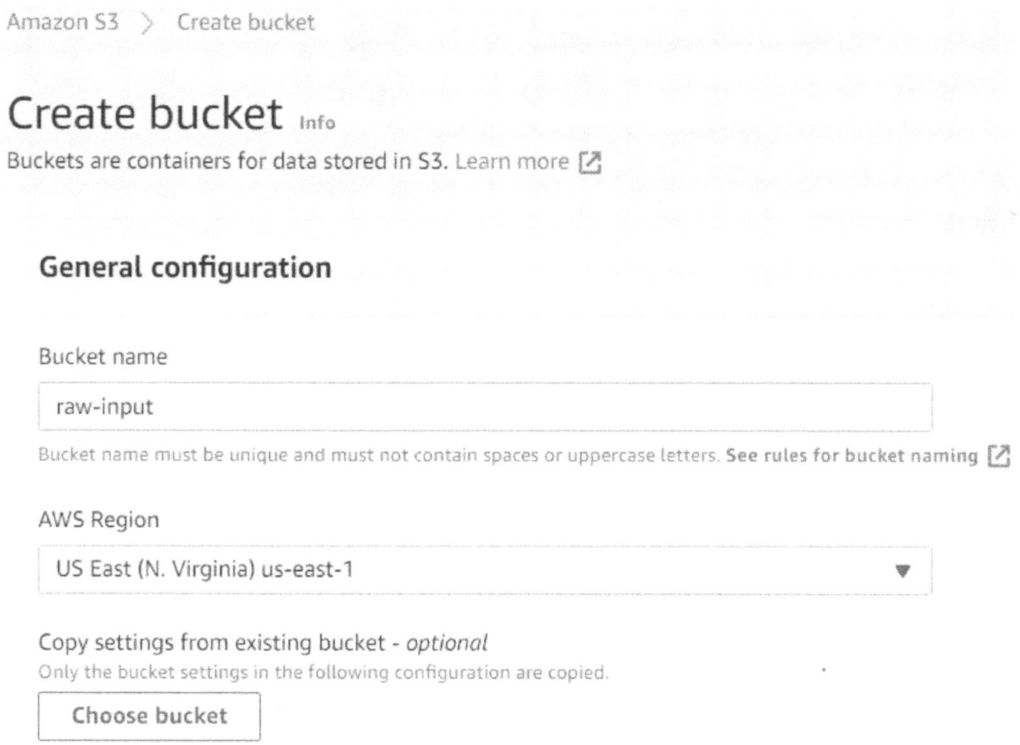

Amazon S3 > Create bucket

Create bucket Info
Buckets are containers for data stored in S3. Learn more [ ]

General configuration

Bucket name

```
raw-input
```
Bucket name must be unique and must not contain spaces or uppercase letters. **See rules for bucket naming** [ ]

AWS Region

```
US East (N. Virginia) us-east-1                                       ▼
```

Copy settings from existing bucket - *optional*
Only the bucket settings in the following configuration are copied.

Choose bucket

Figure 9.2 – Amazon S3 showing the creation of a bucket

After creating the `raw-input` bucket, we can create a subfolder that will be used to capture `sales` data.

The following screenshot shows the creation of the `sales` folder inside the `raw-input` bucket.

Amazon S3 > raw-input > Create folder

Create folder Info

Use folders to group objects in buckets. When you create a folder, S3 creates an object using the name that you specify followed by a slash (/). This object then appears as folder on the console. Learn more

> ⓘ **Your bucket policy might block folder creation**
> If your bucket policy prevents uploading objects without specific tags, metadata, or access control list (ACL) grantees, you will not be able to create a folder using this configuration. Instead, you can use the upload configuration to upload an empty folder and specify the appropriate settings.

Folder

Folder name

| sales | / |

Folder names can't contain "/". See rules for naming

Figure 9.3 – Amazon S3 showing the creation of the sales folder

After the `raw-input` bucket folder structure is created, you can repeat the same step to create a processed output bucket with the `sales` subfolder, which should have the S3 path as `s3://curated-ouput/sales/`.

> **Important Note**
> Please note, Amazon S3 bucket names are globally unique. So, while implementing the solution, you may get an error saying the bucket name already exists. Please provide a unique name and use the same name while implementing the rest of the implementation steps. Forming the bucket name as `<Bucket-Name>-${AWS_ACCOUNT_ID}-${AWS_REGION_CODE}` might help you to get a unique name.

Creating the AWS Lambda function

As explained in the *Architecture overview* section, the objective of the AWS Lambda function is to create an EMR cluster and submit a Spark step that will process the input file.

You can refer to the following steps to create the Lambda function. We have used Python as the language in the Lambda function, but you can integrate your preferred language's equivalent code:

1. Navigate to the AWS Lambda console at `https://console.aws.amazon.com/lambda/home?region=us-east-1`.

2. From the functions list, choose **Create function**, which will open up the form on the web interface, where you can specify **Function name**, **Runtime**, and the IAM role.

3. Then click the **Create function** button to create the function.

The following screenshot shows the **Create function** screen of the AWS console:

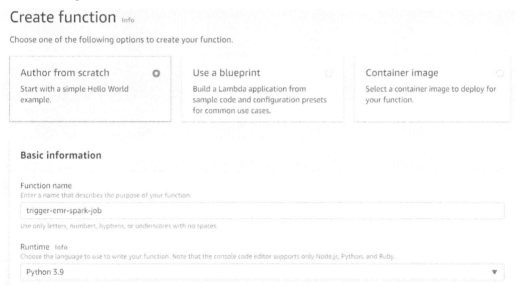

Figure 9.4 – AWS Lambda function creation

After the function is created, you can change its configuration to increase the function timeout, as shown in the following screenshot:

Lambda > Functions > trigger-emr-spark-job > Edit basic settings

Edit basic settings

Basic settings Info

Description - *optional*

Function Timeout

Memory (MB) Info
Your function is allocated CPU proportional to the memory configured.

128   MB

Set memory to between 128 MB and 10240 MB

Timeout

1   min   0   sec

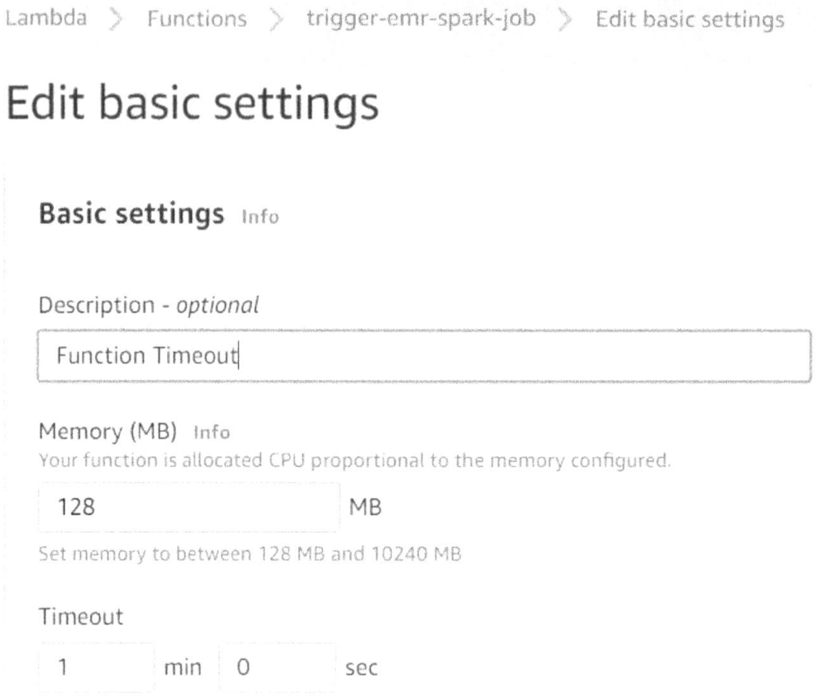

Figure 9.5 – AWS Lambda function – Edit basic settings

After that, you can configure the following three environment variables in the **Configurations** tab and on the **Environment variables** screen:

- **REGION**: This is the AWS region, where the EMR cluster will be created. For our implementation, we have used **us-east-1**.

- **PYSPARK_SCRIPT_PATH**: This will have the Amazon S3 path where you have saved your PySpark script that EMR will execute.

- **S3_OUTPUT_PREFIX**: This is the Amazon S3 path to which the PySpark script will write the output.

Please note, we have not configured the S3 raw input bucket path as that will be passed to the AWS Lambda function dynamically through the **S3 PUT event** when the input CSV file is uploaded into it.

The following screenshot shows the three environment variables configured on the Lambda function:

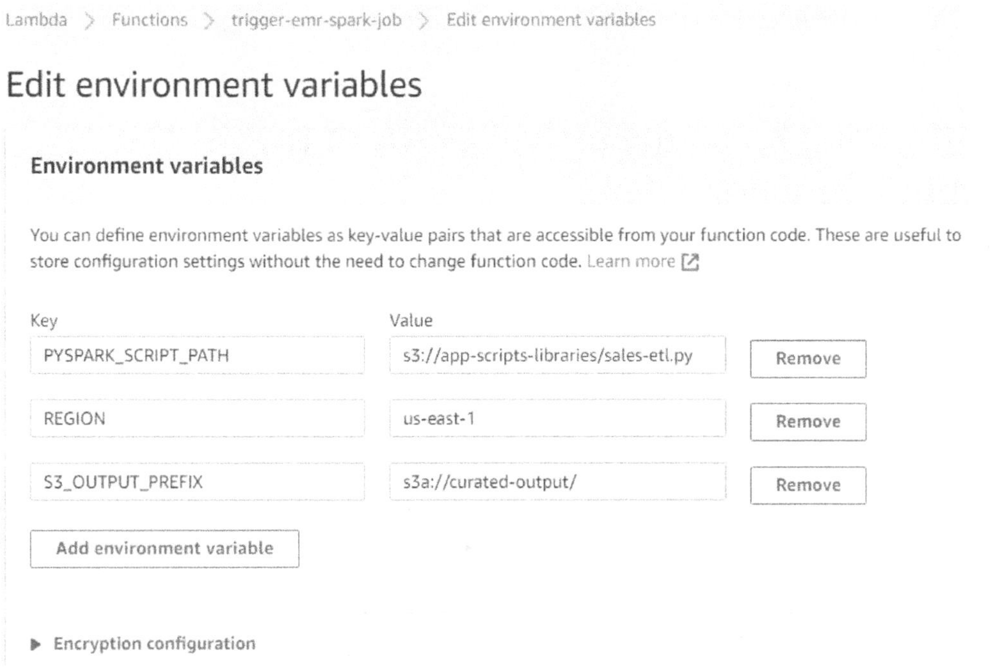

Lambda > Functions > trigger-emr-spark-job > Edit environment variables

Edit environment variables

Environment variables

You can define environment variables as key-value pairs that are accessible from your function code. These are useful to store configuration settings without the need to change function code. Learn more [↗]

Key	Value	
PYSPARK_SCRIPT_PATH	s3://app-scripts-libraries/sales-etl.py	Remove
REGION	us-east-1	Remove
S3_OUTPUT_PREFIX	s3a://curated-output/	Remove

Add environment variable

▶ Encryption configuration

Figure 9.6 – AWS Lambda function – Environment variables

> **Important Note**
> Please make sure the role attached to your Lambda function has permission to invoke EMR APIs and also has permission to write to a CloudWatch log group so that you can debug logs in the event of failures.

As the next step, let's start integrating the Lambda function Python code that invokes the EMR cluster creation API and then adds a Spark execution step. You can get the script from the GitHub repository, specified in the *Technical requirements* section.

Before integrating this Lambda function, make sure you have changed the following variables in the Lambda script as per your environment. In *Chapter 5*, *Setting Up and Configuring Clusters*, we explained cluster creation with advanced options, where we covered the usage of the following cluster parameters:

* `LogUri`
* `Ec2KeyName`
* `Ec2SubnetId`
* `EmrManagedMasterSecurityGroup`
* `EmrManagedSlaveSecurityGroup`

Now that we have integrated the Lambda function, next we can configure it to be triggered with S3 file arrival.

Configuring an S3 file arrival event to trigger the Lambda function

To trigger the Lambda function based on the sales CSV file arrival event, you can refer to the following steps:

1. Navigate to your `raw-input` bucket in Amazon S3.

2. Under the **Properties** tab, navigate to the **Event notifications** section and click **Create event notification**, which will open an event notification form where you can configure the event source, type, and destination.

3. Configure the event source as the `sales/` folder and restrict the event to `.csv` files only, so that if by mistake any other file is uploaded, it does not trigger the event.

The following screenshot of the AWS console shows how you can create an event for a specific folder and file suffix:

General configuration

Event name

trigger-lambda-function

Event name can contain up to 255 characters.

Prefix - *optional*
Limit the notifications to objects with key starting with specified characters.

sales/

Suffix - *optional*
Limit the notifications to objects with key ending with specified characters.

.csv

Figure 9.7 – S3 event notification – General configuration

4. Configure the event type as **All object create events,** as shown in the following screenshot:

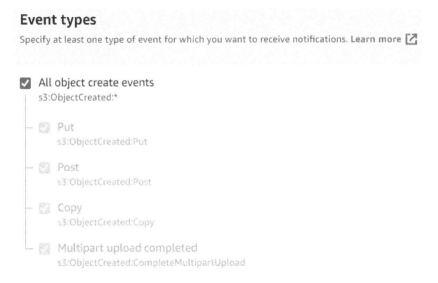

Figure 9.8 – S3 Event notification – Event types

5.   Configure the event destination as the **Lambda function** that we have already created.

Destination

ⓘ Before Amazon S3 can publish messages to a destination, you must grant the Amazon S3 principal the necessary permissions to call the relevant API to publish messages to an SNS topic, an SQS queue, or a Lambda function. Learn more ↗

Destination
Choose a destination to publish the event. **Learn more** ↗

⦿ Lambda function
Run a Lambda function script based on S3 events.

○ SNS topic
Send notifications to email, SMS, or an HTTP endpoint.

○ SQS queue
Send notifications to an SQS queue to be read by a server.

Specify Lambda function

⦿ Choose from your Lambda functions

○ Enter Lambda function ARN

Lambda function

trigger-emr-spark-job	▼

Figure 9.9 – S3 Event notification – Destination

6.   Finally, click **Save** to save the configuration.

After saving, you can confirm the integration by navigating to the AWS Lambda function, which should show the S3 event as a trigger. The following screenshot shows how it should look:

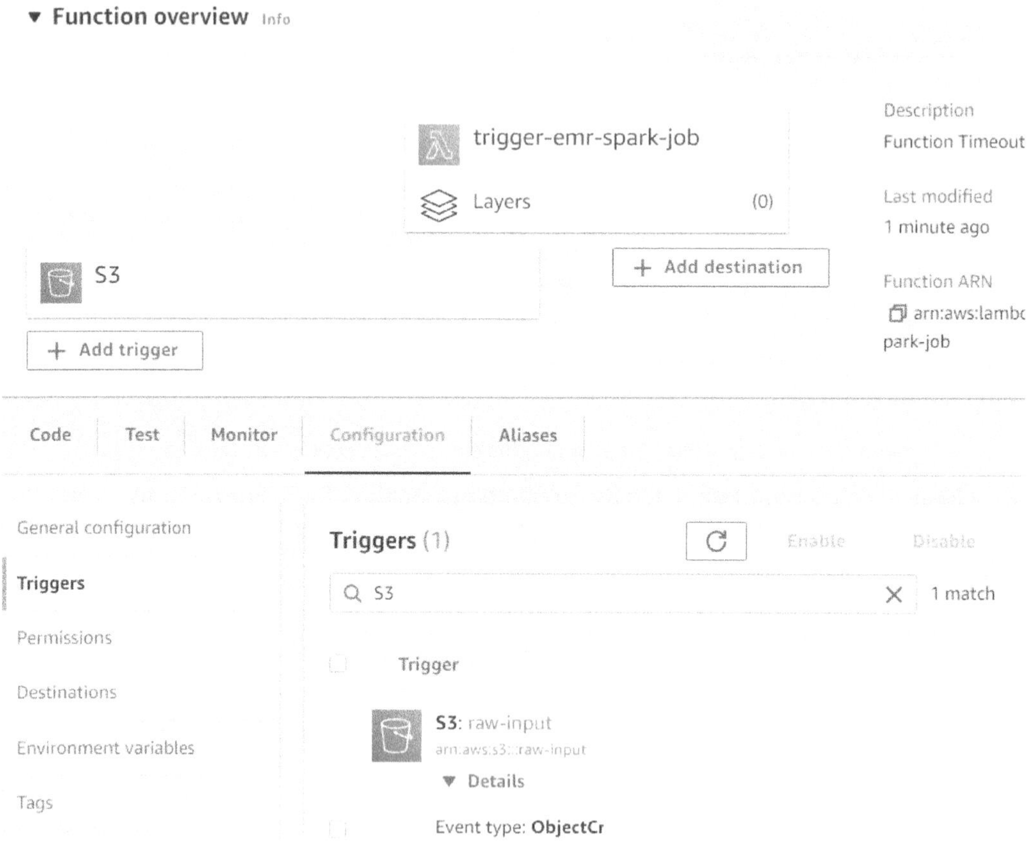

Figure 9.10 – AWS Lambda function – trigger configuration

As the next step, we will add the sales CSV file to the `raw-input` S3 bucket's `sales` folder and see how the Lambda function gets invoked that launches an EMR cluster with the Spark step.

Triggering the EMR job

We have configured an S3 event on the `raw-input` bucket, which will trigger the Lambda function based on the sales CSV file's `PUT` event. To trigger the EMR ETL job, let's add the sample CSV file available at `https://github.com/PacktPublishing/Simplify-Big-Data-Analytics-with-Amazon-EMR-/blob/main/chapter_09/input-data/SalesPipeline_QuickSightSample.csv`, which is a public dataset made available by AWS.

The following screenshot shows that we have uploaded the CSV file to the Amazon S3 bucket using the AWS console:

Figure 9.11 – Amazon S3 shows the CSV file uploaded

After you have uploaded the CSV file, the S3 event should trigger the Lambda function and the Lambda function should launch an EMR cluster with the Spark ETL step.

The following screenshot shows the EMR cluster launch is triggered by Lambda and it's in the **Starting** stage while the Spark ETL job's status is **Pending**.

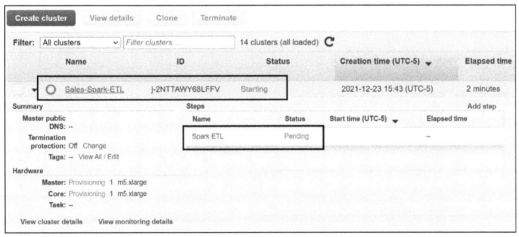

Figure 9.12 – EMR cluster list view that shows EMR cluster resources being launched

After a few minutes, you will notice the cluster resources are provisioned and you can see the Spark job is getting executed. Then, after a few minutes, you will see the Spark job status changes to **Completed** and the cluster gets terminated after job completion. The following screenshot shows the cluster is terminated after job completion:

Figure 9.13 – EMR cluster detail view that shows the job as completed and the cluster as terminated

Next, you can navigate to the Amazon S3 `curated-output` bucket to validate the Spark job has created output in Parquet format and it also created the `<year>/<month>/` partition structure. The following screenshot shows the output written by the Spark job:

Figure 9.14 – S3 output bucket lists Parquet files with a date partition structure

This marks the completion of our ETL process and we can now navigate to Amazon Athena and query the data using SQL.

Validating the output using Amazon Athena

The Parquet format data is already available in Amazon S3 with year and month partition, but to make it more consumable for data analysts or data scientists, it would be great if we could enable querying the data through SQL by making it available as a database table.

To make that integration, we can follow a two-step approach:

1. We can run the Glue crawler to create a Glue Data Catalog table on top of the S3 data.

2. We can run a query in Athena to validate the output.

Let's see how you can integrate that.

Defining a virtual Glue Data Catalog table on top of Amazon S3 data

You can follow these steps to create and run the Glue crawler, which will create a Glue Data Catalog table:

1. Navigate to the AWS Glue crawler at `https://console.aws.amazon.com/glue/home?region=us-east-1#catalog:tab=crawlers`.

2. Then click **Add crawler**, which will open up the form to configure the crawler.

3. Configure the crawler, where the data source should point to the `curated-output` S3 bucket.

4. Specify the IAM role that has permission to crawl the S3 bucket.

5. You can keep the rest of the configurations as the defaults and then, on the final screen, review the configurations that can look like *Figure 9.15* and click **Save**.

6. Select the crawler you created using the previous steps from the crawler list and select the **Run crawler** button, which will create a table in Glue Data Catalog with the name `curated_output`.

The following screenshot shows the Glue crawler's review page, which shows the configurations we have specified for it:

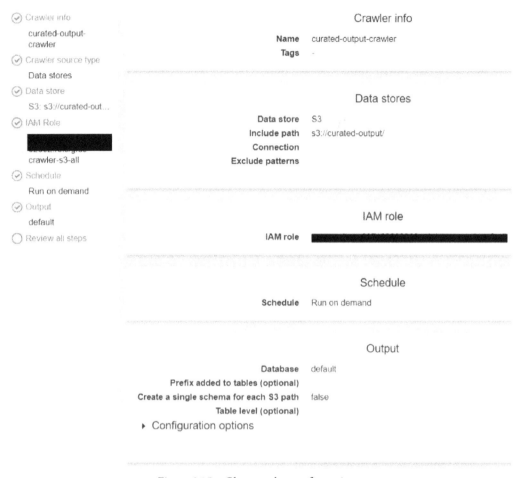

Figure 9.15 – Glue crawler configuration

Now that we have the Glue Data Catalog table created, we can navigate to Amazon Athena to query the data using SQL.

Querying output data using Amazon Athena standard SQL

In Athena, you can keep **Data Source** as the default **AwsDataCatalog** and select **default** for **Database**. Then execute the following SQL query to validate the output:

```
SELECT * FROM "default"."curated_output" LIMIT 10;
```

The following screenshot shows the Athena query execution output.

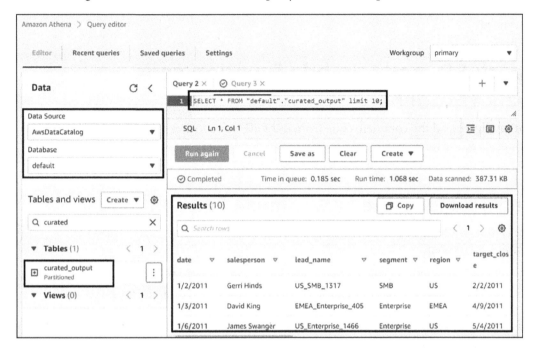

Figure 9.16 – Athena Query editor with the query result

Please note, if you are going to use Amazon Athena for the first time, then AWS expects you to set up a query result location in Amazon S3. You can click on **View Settings** and then click on **Manage** to provide the S3 path.

The following screenshot shows where you can configure it:

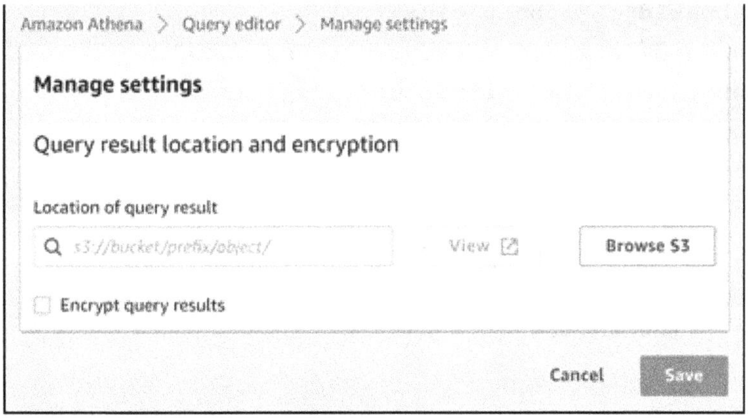

Figure 9.17 – Athena Query editor settings to specify the S3 path for query results

This concludes our use case implementation and validation steps. Next, we will walk through the AWS Lambda and PySpark script so that you can modify them as per your need.

Spark ETL and Lambda function code walk-through

You can download the complete code from our GitHub repository specified in the *Technical requirements* section of the chapter. In this section, we will highlight a few sections of the code to explain its purpose and usage.

Understanding the AWS Lambda function code

The Lambda function's primary objective is to invoke EMR cluster launch and then submit a Spark step.

The following part of the code creates a `boto3` client for the EMR service and invokes the `run_job_flow` method of it such that it takes all the required inputs for the cluster:

```
conn = boto3.client("emr", region_name=AWS_REGION)
cluster_id = conn.run_job_flow(…)
```

The following parameters are passed to the `run_job_flow` method that specifies the EMR cluster configurations:

```
Instances={
    "Ec2KeyName": "<key-name>",
    "Ec2SubnetId": "subnet-<id>",
    "EmrManagedMasterSecurityGroup": "sg-<id>",
    "EmrManagedSlaveSecurityGroup": "sg-<id>",
    "HadoopVersion": "Amazon 3.2.1",
    'InstanceGroups': [
        {
            'Name': 'Master nodes',
            'Market': 'ON_DEMAND',
            'InstanceRole': 'MASTER',
```

```
            'InstanceType': 'm5.xlarge',
            'InstanceCount': 1,
        },
        {
            'Name': 'Slave nodes',
            'Market': 'ON_DEMAND',
            'InstanceRole': 'CORE',
            'InstanceType': 'm5.xlarge',
            'InstanceCount': 1,
        }
    ],
    'KeepJobFlowAliveWhenNoSteps': False,
    'TerminationProtected': False
}
```

The following part of the script specifies the Spark step with the parameters required for the `spark-submit` command:

```
Steps=[
    {
        'Name': 'Spark ETL',
        'ActionOnFailure': 'TERMINATE_CLUSTER',
        'HadoopJarStep': {
            'Jar': 'command-runner.jar',
            'Args': [
                "spark-submit", "--deploy-mode", "cluster",
                "--master", "yarn",
                PYSPARK_SCRIPT_PATH, S3_INPUT_PATH, S3_OUTPUT_
PREFIX+currentYear+"/"+currentMonth+"/"
            ]
    }
}]
```

Now, let's understand how the PySpark ETL code is integrated.

Understanding the PySpark script integrated into the EMR step

You can download the complete PySpark script from our GitHub repository specified in the *Technical requirements* section, but the following part of the script is the core of the script. This code block shows how you can read the input file with the `spark.read` method, which returns a Spark DataFrame, apply transformations to replace column names, and then write processed data to the output path with the `<dataframe>.write.parquet` method:

```
df = spark.read.format("csv").option("header", "true").
load(sys.argv[1])
replacements = {c:c.replace(' ','_') for c in df.columns if ' '
in c}
df1 = df.select([col(c).alias(replacements.get(c, c)) for c in
df.columns])
df1.write.parquet(sys.argv[2])
```

Here, `argv[1]` will provide the S3 input path and `argv[2]` will provide the S3 output path. This script should work without any modification, but you can customize it as per your ETL transformation logic.

Summary

Over the course of this chapter, we have dived deep into a batch ETL use case, where we integrated the data pipeline with Amazon S3, AWS Lambda, Amazon EMR, AWS Glue, and Amazon Athena.

We have covered detailed implementation steps, which you can follow to replicate the steps or customize them as per your use case.

At the end of the chapter, we provided an overview of a few important parts of the AWS Lambda function and EMR PySpark script, which can provide you with a starting point for your projects.

That concludes this chapter! Hopefully, this helped you get an idea of how batch ETL pipelines can be integrated, and in the next chapter, we will integrate another use case, which is real-time streaming with Amazon EMR.

Test your knowledge

Before moving on to the next chapter, test your knowledge with the following questions:

1. Assume you have integrated the complete ETL pipeline but when your input file gets pushed to the input S3 bucket, the Lambda function does not launch the EMR cluster. When you plan to debug the Lambda function execution, you don't find any logs for the Lambda function in CloudWatch log groups. What might be the problem that stops the Lambda function from writing logs in CloudWatch and how would you resolve it?

2. Assume you have multiple data sources that are sending input files for processing. Instead of triggering an EMR cluster launch on an S3 file arrival event, you would like to schedule a PySpark job to run at a particular time of the day, so that it picks up all the input files available at that point of time for processing. How would you schedule the cluster creation and job execution?

3. You have integrated Amazon EMR for your batch analytics workload that is scheduled to run every day at midnight. But occasionally you notice that the job completion takes more time than expected and on a few occasions, the jobs fail too. How would you build a retry mechanism for the job failures and also a notification feature that notifies administrators of job failures?

Further reading

Following are a few resources you can refer to for further reading:

*   Lambda function for transient EMR cluster use cases: `https://docs.aws.amazon.com/prescriptive-guidance/latest/patterns/launch-a-spark-job-in-a-transient-emr-cluster-using-a-lambda-function.html`

*   More on AWS Glue crawler definition and execution: `https://docs.aws.amazon.com/glue/latest/dg/add-crawler.html`

*   Optimize Spark performance in EMR: `https://docs.aws.amazon.com/emr/latest/ReleaseGuide/emr-spark-performance.html`

10

Implementing Real-Time Streaming with Amazon EMR and Spark Streaming

In *Chapter 3*, *Common Use Cases and Architecture Patterns*, we discussed different use cases and architecture patterns that you can follow using Amazon EMR, while in *Chapter 9*, *Implementing Batch ETL Pipeline with Amazon EMR and Apache Spark*, you learned how you can implement a batch **Extract, Transform, and Load** (ETL) pipeline using **Amazon EMR** and **PySpark** script.

In this chapter, we will dive deep into another use case – **real-time streaming with Amazon EMR and Spark Streaming**, where we will look at the implementation steps that you can follow to replicate the setup in your AWS account.

Real-time streaming use cases are becoming more popular as distributed processing engines such as Spark can stream, transform in real time, and help drive business decisions through real-time **business intelligence** (**BI**) reporting. This sample use case implementation learning will provide you with a starting point from where you can build a more complex, real-time data pipeline in AWS using Amazon EMR.

We will cover the following topics, which will help you understand the use case, its application architecture, and how a transient EMR cluster with Spark can be integrated for distributed processing:

- Use case and architecture overview

- Implementation steps

- Validating output using Amazon Athena

- Spark streaming code walk-through

Technical requirements

In this chapter, we will implement a real-time streaming pipeline using AWS analytics services. So, before getting started, you need to make sure that you have the following requirements ready:

- An AWS account with access to create Amazon S3, Amazon EMR, Amazon Athena, Amazon Cognito, and AWS Glue Catalog resources.

- An IAM user who has access to create IAM roles, which will be used to trigger AWS CloudFormation stack or execute jobs.

Refer to the following link for access to the book's GitHub repository: `https://github.com/PacktPublishing/Simplify-Big-Data-Analytics-with-Amazon-EMR-/tree/main/chapter_10`.

Now, let's dive deep into the use case and the hands-on implementation steps.

Check out the following video to see the Code in Action at `https://bit.ly/3oIz89Q`

Use case and architecture overview

For this use case, let's assume you have a consumer-facing website where users are interacting with your web pages by clicking different buttons or links, which are specific to different page navigations, signing up, signing in, or buying products. You have started a promotional sale on your website for a limited duration and you would like to track how users are reacting to it in real time.

To track user activity, your frontend application has integrated click events, which will publish a JSON event string with every mouse click to a **message bus** such as **Amazon Kinesis Data Streams** or **Kafka**. From the message bus, a **Spark Streaming**-based consumer application will read JSON messages as a micro-batch and write to Amazon S3 data lake for real-time analysis.

To replicate the streaming of click events, we will integrate the **Kinesis Data Generator** web UI tool, where you can configure a sample JSON event and schedule it to publish a fixed number of records to the Kinesis Data Streams message bus. Then, we will leverage Amazon EMR with Spark Streaming as the stream consumer application. This EMR cluster will be a persistent EMR cluster that is always active with a minimal number of nodes and with autoscaling built in to scale resources as the data volume grows.

Now, let's get an overview of the architecture.

Architecture overview

The following is the high-level architecture diagram of the streaming pipeline:

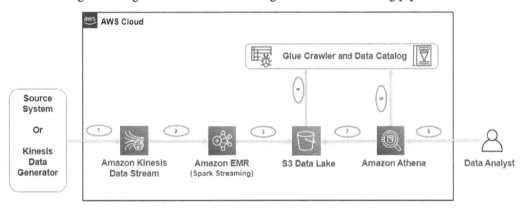

Figure 10.1 – Reference architecture diagram for a real-time streaming pipeline

Let's take a closer look at this diagram:

- *Step #1* represents the source application publishing JSON click events to Kinesis Data Streams. In our implementation, we will leverage Kinesis Data Generator to simulate the streaming data ingestion.

- *Step #2* represents an Amazon EMR cluster with a Spark Streaming job reading messages from Kinesis Data Streams, and *step #3* represents the Spark job writing Parquet format output to the Amazon S3 data lake with /<year>/<month>/ partition columns.

- *Step #4* represents Glue Catalog tables defined on top of the data lake bucket, which will be used by Amazon Athena for querying.

- *Step #5*, *#6*, and *#7* represent data analysts using Amazon Athena for querying data using standard SQL. Amazon Athena queries table metadata from the Glue Catalog and then queries data from Amazon S3 to show the output in a structured tabular format to the end user.

After getting an overview of the use case and architecture, let's get started on the implementation steps. Please make sure that you meet all the prerequisites defined in the *Technical requirements* section of this chapter.

Implementation steps

In this section, we will guide you through the implementation steps for the use case and architecture we explained in the previous section.

> **Important Note**
>
> While explaining the implementation steps, we have used **us-east-1** as the AWS region. You can use the same or an alternate region as per your choice. Please check any resource or service limits that might apply to your AWS region before proceeding with the implementation.

Creating Amazon S3 buckets

Let's first create the Amazon S3 buckets, which will be used by the EMR Spark job to write the streaming data. Please refer to the following steps to create them:

1. Navigate to the Amazon S3 console at `https://s3.console.aws.amazon.com/s3/home?region=us-east-1#`.

2. From the buckets list, choose the **Create bucket** option, which will open a form on the web interface to provide your bucket name and related configurations.

   We have specified the bucket name as `clickstream-events` and kept everything else at their default settings.

3. Then, click the **Create bucket** button to create the bucket.

The following screenshot shows the AWS console, through which we have created the bucket:

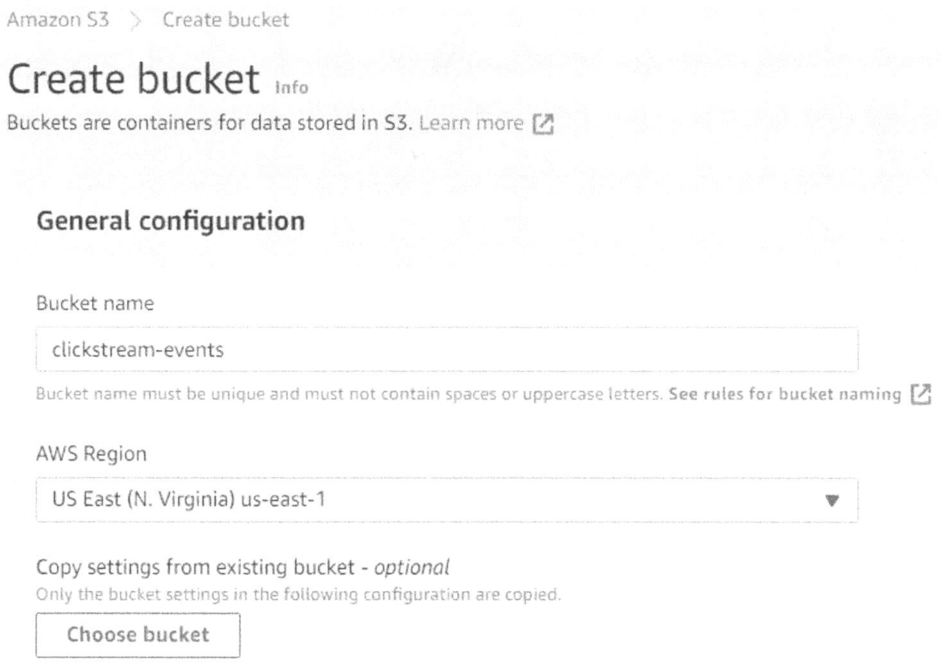

Amazon S3 > Create bucket

Create bucket Info

Buckets are containers for data stored in S3. Learn more ↗

General configuration

Bucket name

clickstream-events

Bucket name must be unique and must not contain spaces or uppercase letters. **See rules for bucket naming** ↗

AWS Region

US East (N. Virginia) us-east-1   ▼

Copy settings from existing bucket - *optional*
Only the bucket settings in the following configuration are copied.

Choose bucket

Figure 10.2 – Creation of a bucket in Amazon S3

As explained in the previous chapter, you may not be able to use the same bucket name as the S3 buckets are globally unique. Please make sure to provide a valid S3 bucket name that is unique and use the same name for the remainder of the implementation.

After creating the S3 bucket, next, we will create the Kinesis data stream and the EMR cluster.

Creating the Amazon Kinesis data stream

As explained in the architecture overview section, the objective of integrating an Amazon Kinesis data stream is to create an aggregator layer that can receive stream events from multiple producer applications and can have multiple consumer applications reading from the stream. In our use case, we just have one producer, which is the Kinesis Data Generator tool, and one consumer, which is the EMR Spark application.

You can refer to the following steps to create the Kinesis data stream:

1. Navigate to the Amazon Kinesis Data Streams console at `https://console.aws.amazon.com/kinesis/home?region=us-east-1#/home`.

2. You will notice that **Kinesis Data Streams** is selected by default, and you can click **Create data stream**, which will open the **Create data stream** form.

3. You can specify the stream name and leave everything else as their default settings and then click **Create data stream** to create the stream. For this implementation, we have taken a very small cluster with just one shard. A shard in a Kinesis data stream is the basis of the throughput unit. Depending on the amount of data you are writing or reading from the stream, you can size your cluster with the required number of shards.

The following screenshot shows the **Create data stream** screen of the AWS console:

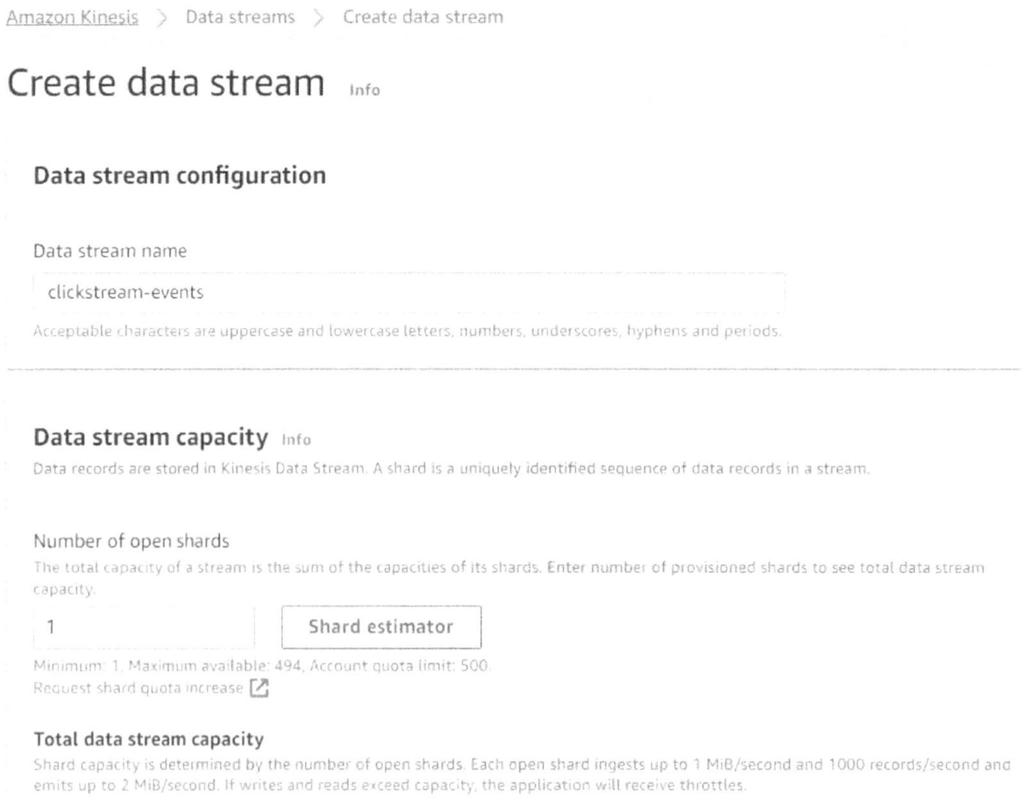

Amazon Kinesis > Data streams > Create data stream

Create data stream   Info

Data stream configuration

Data stream name

clickstream-events

Acceptable characters are uppercase and lowercase letters, numbers, underscores, hyphens and periods.

Data stream capacity   Info
Data records are stored in Kinesis Data Stream. A shard is a uniquely identified sequence of data records in a stream.

Number of open shards
The total capacity of a stream is the sum of the capacities of its shards. Enter number of provisioned shards to see total data stream capacity.

| 1 | Shard estimator |

Minimum: 1. Maximum available: 494. Account quota limit: 500.
Request shard quota increase

Total data stream capacity
Shard capacity is determined by the number of open shards. Each open shard ingests up to 1 MiB/second and 1000 records/second and emits up to 2 MiB/second. If writes and reads exceed capacity, the application will receive throttles.

Figure 10.3 – Amazon Kinesis Data Streams – Create data stream

After the Kinesis data stream cluster is created, next we can set up the Kinesis Data Generator tool and configure it to publish sample JSON events to Kinesis Data Streams.

Creating and configuring the Kinesis Data Generator tool

The Kinesis Data Generator tool is an open source tool available in GitHub that provides a web interface through which you can publish sample events to Kinesis Data Streams or Kinesis Data Firehose. While configuring the tool to publish events, you can create a reusable template and use that to publish thousands of events per second.

The tool also needs an **Amazon Cognito** user pool to be created, which you will use to log in to the Kinesis Data Generator tool. To do the complete setup, there is a **CloudFormation** template that helps to create the required resources for the tool.

Amazon Cognito is a serverless scalable service that lets you manage your user's sign-up and sign in methods and can be easily integrated with your web and mobile applications. It also provides a sign-in mechanism with different social identity providers such as Facebook and Google.

AWS CloudFormation provides you with the capability to build automated DevOps pipelines, where you can create and orchestrate infrastructure resources using code. The Kinesis Data Generator tool is set up using a CloudFormation template where it is written to create all the required resources, including the Amazon Cognito user who will be used for logging in.

To set up the tool, please refer to the following steps:

1. Navigate to the Kinesis Data Generator tool help page at `https://awslabs.github.io/amazon-kinesis-data-generator/web/help.html`.

2. Click the **Create a Cognito User with CloudFormation** button, which will take you to the AWS Cloud Formation's **Create stack** screen with **Template source** already populated with the template S3 path. You need to make sure you are still in the **us-east-1** region, otherwise, you can change the region to **us-east-1**. The following represents a screenshot of the CloudFormation screen:

Figure 10.4 – CloudFormation Create stack screen

3. Then, click **Next**.

    On the next screen, you can populate the **Stack name**, **Username**, and **Password** fields for the Cognito user who will be newly created by the CloudFormation stack. The following screenshot shows the next screen view:

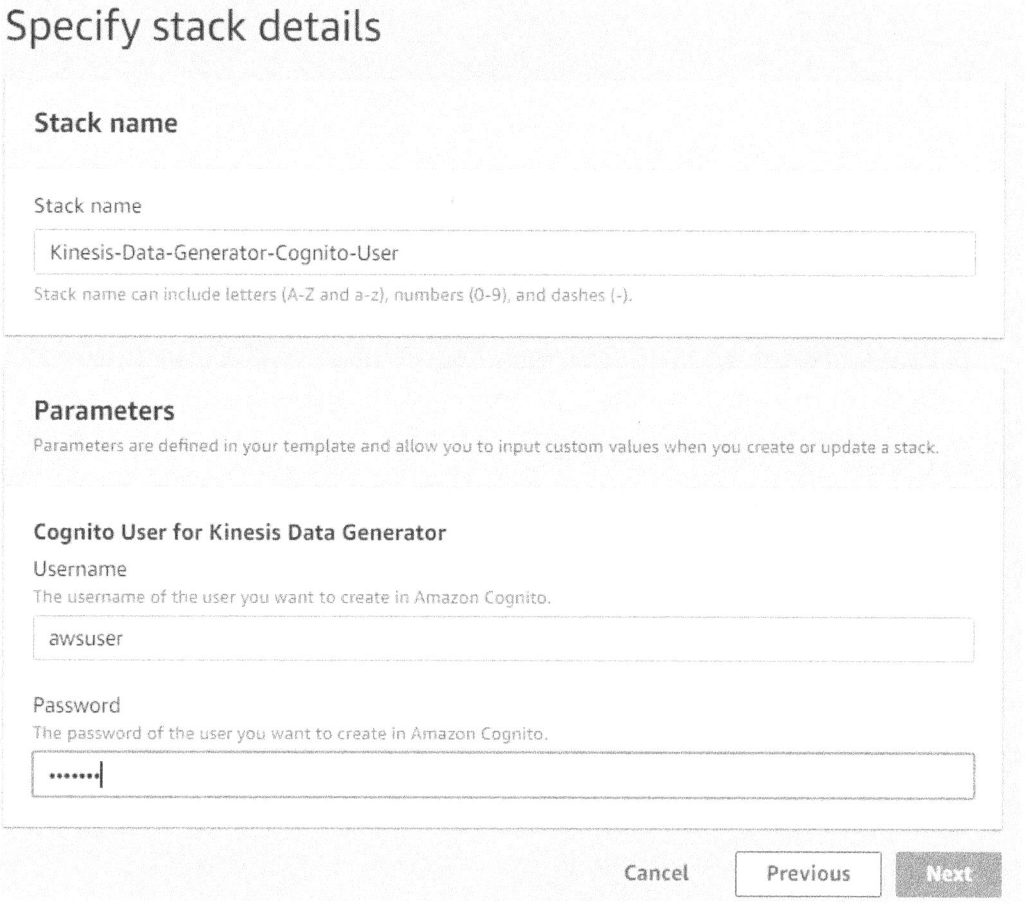

Specify stack details

Stack name

Stack name

Kinesis-Data-Generator-Cognito-User

Stack name can include letters (A-Z and a-z), numbers (0-9), and dashes (-).

Parameters

Parameters are defined in your template and allow you to input custom values when you create or update a stack.

Cognito User for Kinesis Data Generator

Username
The username of the user you want to create in Amazon Cognito.

awsuser

Password
The password of the user you want to create in Amazon Cognito.

•••••••

Cancel   Previous   Next

Figure 10.5 – CloudFormation screen that shows the stack parameters

4. On the final screen, click **Create stack**, which will create all the required resources.

5.   After the stack status changes to **CREATE_COMPLETE**, you can navigate to the **Outputs** tab and click the URL specified for the **KinesisDataGeneratorUrl** key.

Figure 10.6 – CloudFormation Outputs tab

6.   This will take you to the Kinesis Data Generator UI, where you can specify your Cognito user credentials to log in. The following screenshot shows the login screen:

Amazon Kinesis Data Generator

The KDG makes it simple to send test data to your Amazon Kinesis stream or Amazon Kinesis Firehose delivery stream. Sign in to get started. If you haven't configured an Amazon Cognito user, choose Help.

Figure 10.7 – Kinesis Data Generator tool's login

This completes the Kinesis Data Generator tool setup. As a next step, we can configure the sample JSON click events that we plan to publish to Kinesis Data Streams.

Configuring Kinesis Data Generator to publish JSON events to Kinesis Data Streams

As a next step, we need to configure the tool to publish JSON events to the Kinesis data stream that we have created.

Here is the sample JSON event that we plan to publish:

```
{"browser": "{{random.arrayElement(["Chrome","Safari","IE",
"Edge"])}}", "device": "{{random.
arrayElement(["Mobile","Desktop"])}}", "platform": "{{random.
arrayElement(["Win10","iOS","macOS"])}}", "referer": "{{random.
arrayElement(["www.google.com","www.yahoo.com","www.aol.
com","www.amazon.com"])}}", "request_time": "{{date.now}}",
"user_address": "{{internet.ip}}"}
```

This will generate random values for the `browser`, `device`, `platform`, and `referer` attributes from a pre-defined set of string values and will generate dynamic values for the `request_time` and `user_address` attributes.

To configure the tool to publish these events as a continuous stream, please refer to the following steps:

1. After navigating to the Kinesis Data Generator tool in the previous section, you will see a form where you need to specify the AWS region, select the Kinesis data stream name, and also specify the input JSON that you plan to publish to Kinesis Data Streams.

    After populating the values, it should look like the following screen:

Figure 10.8 – Kinesis Data Generator tool screen with the JSON event

2. Then, click **Send data**, which will publish JSON events to your stream at a default rate of 100 records per second iteratively. This can be changed in the **Records per second** form field before submitting. The publishing of events is shown on an overlay popup with the **Stop Sending Data to Kinesis** button, as shown in the following screenshot:

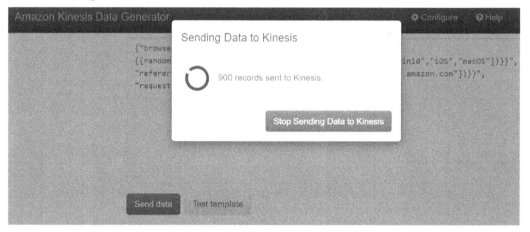

Figure 10.9 – Kinesis Data Generator tool screen to stop sending JSON events

Don't click the **Stop Sending Data to Kinesis** button yet as we plan to integrate the Spark Streaming application to read from the stream in real time. Next, you can validate whether Kinesis Data Streams is receiving the events.

Validating the input data in Kinesis Data Streams

To validate the data ingestion in Kinesis Data Streams, please refer to the following steps:

1. Navigate to the Amazon Kinesis Data Streams console at `https://console.aws.amazon.com/kinesis/home?region=us-east-1#/home`.

2. Click **Data streams** from the left navigation, which will list all the Kinesis data streams you have created.

3. Click the `clickstream-events` stream that we created in the previous step and navigate to the **Monitoring** tab. The following charts on the monitoring tab will show the data ingested so far into the stream.

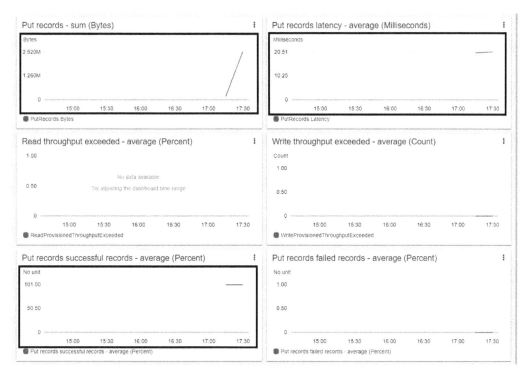

Figure 10.10 – Kinesis Data Streams console – Stream monitoring tab

As you can see from the charts, it confirms the data being ingested through the Kinesis Data Generator UI tool. Next, let's try to set up the Amazon EMR cluster and integrate a Spark Streaming application.

Creating an Amazon EMR cluster and configuring a Spark Streaming job

To create an EMR cluster, refer to the following steps. These steps are the same as we discussed in *Chapter 5, Setting Up and Configuring EMR Clusters:*

1. Navigate to Amazon EMR's **Create cluster** screen at `https://console.aws.amazon.com/elasticmapreduce/home?region=us-east-1#quick-create`.

2. Specify **Cluster name** and select one of the latest stable **Release** versions. We have selected the **emr-6.4.0** release as that was the latest stable release at the time of writing this chapter. From the **Applications** list, select the Spark application stack as we plan to integrate a Spark Streaming job.

The following is a screenshot of the values we have selected, which can guide you:

Figure 10.11 – Amazon EMR Create cluster screen – General and Software configurations

3. Next, under **Hardware configuration**, enable **Cluster scaling** with default values for **EMR-managed scaling**. Then, under **Security and access**, select your **EC2 key pair** with IAM roles pointing to the default roles. The following screenshot shows the values we have selected:

Hardware configuration

Instance type   [m5 xlarge   ∨]   The selected instance type adds 64 GiB of GP2 EBS
                                                                        storage per instance by default. Learn more ⎗

Number of instances   [3]   (1 master and 2 core nodes)

Cluster scaling   ☑ scale cluster nodes based on workload

                   EMR-managed scaling

                   EMR will automatically increase and decrease the number of instances in core and task nodes based on
                   workload. Set a minimum and maximum limit of the number of instances for the cluster nodes. Master nodes
                   do not scale. Learn more ⎗

Core and task units

Minimum: [2] ⬍

Maximum: [10] ⬍

Auto-termination   ☐ Enable auto-termination   Learn more ⎗

Security and access

EC2 key pair   [███████████████   ∨]   ⓘ Learn how to create an EC2 key pair

Permissions   ⦿ Default   ○ Custom

               Use default IAM roles. If roles are not present, they will be automatically created
               for you with managed policies for automatic policy updates.

EMR role   EMR_DefaultRole ⎗   ○ Use EMR_DefaultRole_V2 ⓘ

EC2 instance profile   EMR_EC2_DefaultRole ⎗ ⓘ

Cancel   [Create cluster]

Figure 10.12 – Amazon EMR Create cluster screen – Hardware and Security and access configurations

4. Then, select **Create cluster**, which will take you to the EMR cluster detail screen
    with a status of **Starting**.

After a few minutes, you will notice that the cluster status changes to **Running** when the
initial **Setup hadoop debugging** default job runs, and then, following the completion of
the job, it changes to **Waiting**, which means all the resources are provisioned and we are
good to submit jobs to the cluster.

Now, let's see how we can trigger the Spark Streaming job.

Triggering the Spark Streaming job on the EMR cluster

As a next step, we need to trigger the Spark Streaming job by adding it as a step in the EMR cluster. To add the step, click the **Steps** tab of the cluster detail page and click the **Add Step** button, which will open the following screen to add the step:

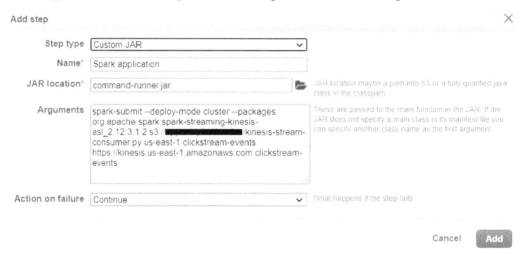

Figure 10.13 – Amazon EMR – Add step

You can select **Custom JAR** in the **Step type** field and specify the following spark-submit command, which triggers the kinesis-stream-consumer.py Spark script available in an S3 bucket:

```
spark-submit --deploy-mode cluster --packages org.apache.
spark:spark-streaming-kinesis-asl_2.12:3.1.2 s3://<script-
bucket-name>/kinesis-stream-consumer.py us-east-1 clickstream-
events https://kinesis.us-east-1.amazonaws.com clickstream-
events
```

The following is the format of the command:

```
spark-submit --deploy-mode cluster --packages org.apache.
spark:spark-streaming-kinesis-asl_2.12:3.1.2 s3://<script-
bucket-name>/<script-name>.py <aws-region> <kinesis-data-
stream-name> <kinesis-endpoint-url> <s3-output-bucket-name>
```

Please replace the `<script-bucket-name>`, `<script-name>`, `<aws-region>`, `<kinesis-data-stream-name>`, `<kinesis-endpoint-url>`, and `<s3-output-bucket-name>` variables before adding the EMR step. You can find the Kinesis endpoint URL for your AWS region in the AWS documentation.

> **Important Note**
>
> Please note that the `spark-submit` command has an additional `--packages` parameter, which specifies the `spark-streaming-kinesis-asl` Maven JAR path. The 3.1.2 version of the JAR is integrated, based on the Spark version available in EMR 6.4.0, so please change the version if you are executing the command on a different EMR cluster release version.

After you have added the step, you will notice that the step status changes to **Running** and, after a few minutes, you will start seeing records in your target S3 bucket, as follows:

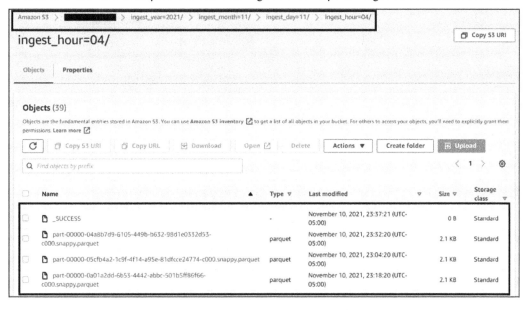

Figure 10.14 – Amazon S3 screen representing Parquet output

As you can see, the Parquet files are being written to the `ingest_year, ingest_month, ingest_day,` and `ingest_hour` partition columns.

You can increase the number of records in the Kinesis Data Generator UI tool and can validate how the EMR manages scaling and increases the number of core nodes in the EMR cluster to handle the load. The following screenshot shows the node status while additional nodes are being provisioned.

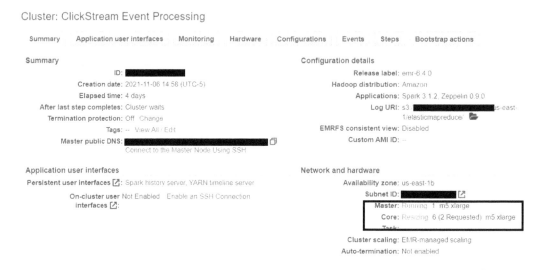

Figure 10.15 – Amazon EMR cluster detail screen showing core node resizing

This concludes our implementation steps. Next, we can validate the data by querying it using Amazon Athena.

Validating output using Amazon Athena

The Parquet format data is already available in Amazon S3 partition columns, but to make it more consumable for data analysts or data scientists, it would be great if we can enable querying the data through SQL by making it available as a database table.

To make that integration, we will follow a two-step approach:

1. First, we will run Glue Crawler to create a Glue Catalog table on top of the S3 data.

2. Then, we will run a query in Athena to validate the output.

Let's see how you can integrate that.

Defining a virtual Glue Catalog table on top of Amazon S3 data

You can follow these steps to create and run Glue Crawler, which will create a Glue Data Catalog table:

1. Navigate to AWS Glue Crawler at `https://console.aws.amazon.com/glue/home?region=us-east-1#catalog:tab=crawlers`.

2. Then, click **Add crawler**, which will open a form to configure the crawler.

3. Configure the crawler, where the data source should point to the `clickstream-events` S3 bucket.

4. Specify the IAM role that has permission to crawl the S3 bucket.

5. You can keep the rest of the configurations at their default settings and then, on the final screen, review the configurations that might look like the *Figure 10.16* and click **Save**.

6. Select the crawler you created using the preceding steps from the crawler list and then select the **Run crawler** button, which will create a table in the Glue Catalog with a name of `clickstream_events`.

After we have the Glue Catalog table created, we can now navigate to Amazon Athena to query the data using SQL.

Querying output data using a standard SQL query in Amazon Athena

To validate the output, navigate to Amazon Athena at `https://console.aws.amazon.com/athena/home?region=us-east-1`.

As explained in *Chapter 9, Implementing Batch ETL Pipeline with Amazon EMR and Apache Spark*, if you are accessing Amazon Athena for the first time, configure the Athena query result location by pointing it to an Amazon S3 path.

In the Athena query editor, you can keep the **Data Source** field as the default **AwsDataCatalog** and select **default** for the database. Then, execute a SQL query such as the following to validate the output:

```
SELECT * FROM "default"."clickstream_events" limit 10;
```

The following screenshot shows Amazon Athena output for the SQL query:

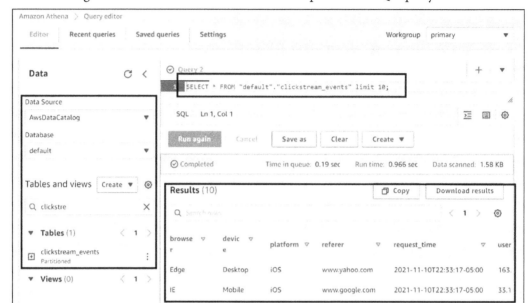

Figure 10.16 – Athena query editor with a query result

This concludes our use case implementation and validation steps. Next, we will look at a walk-through of the PySpark script so that you can modify it as per your requirements.

> **Important Note**
>
> After validating the data using Amazon Athena, please navigate to the Kinesis Generator UI tool and make sure you stop sending records to avoid incurring costs.

Now, let's learn about Spark Streaming code in the following section.

Spark Streaming code walk-through

You can download the complete PySpark script from our GitHub repository. The following is a walk-through of the primary functions of the script.

The following `getSparkSessionInstance()` function is a user-defined function that gets an existing SparkSession, instead of creating a duplicate instance within custom, user-defined functions:

```
# Get existing SparkSession
def getSparkSessionInstance(sparkConf):
```

```
  if ("sparkSessionSingletonInstance" not in globals()):
      globals()["sparkSessionSingletonInstance"] = SparkSession.
builder.config(conf=sparkConf).getOrCreate()
  return globals()["sparkSessionSingletonInstance"]
```

The following `processRecords()` function is a user-defined function, which is being invoked by each RDD of the Kinesis stream to parse the records of the RDD and write to Amazon S3 in Parquet format with `year`, `month`, `date`, and `hour` partition columns:

```
# Process each RDD and write to S3
def processRecords(rdd):
  if not rdd.isEmpty():
    spark = getSparkSessionInstance(rdd.context.getConf())
    df = spark.read.json(rdd)
    now = datetime.datetime.now()
    year = now.strftime("%Y")
    month = now.strftime("%m")
    day = now.strftime("%d")
    hour = now.strftime("%H")
    TargetPath = "s3://"+s3OutputPath+"/ingest_year="+ year
+ "/ingest_month=" + month + "/ingest_day=" + day + "/ingest_
hour=" + hour + "/"
    df.write.mode('append').parquet(TargetPath)
```

The following is the primary code, which initiates `StreamingContext`, reads from the Kinesis data stream, and invokes the preceding `processRecords()` function for each RDD function:

```
# Read Runtime arguments
scriptName, regionName, streamName, endpointUrl, s3OutputPath =
sys.argv


# Initialize SparkSession and StreamingContext
appName = "ClickStreamEventConsumer"
sparkSession =  SparkSession.builder.appName(appName).
getOrCreate()
sparkContext = sparkSession.sparkContext
streamingContext = StreamingContext(sparkContext, 1)
```

```
# Start reading from Kinesis Stream
records = KinesisUtils.createStream(streamingContext, appName,
streamName, endpointUrl, regionName, InitialPositionInStream.
LATEST, 5, StorageLevel.MEMORY_AND_DISK_2)


# Process each batch
records.foreachRDD(lambda rdd: processRecords(rdd))


# Start StreamingContext
streamingContext.start()
streamingContext.awaitTermination()
```

Here, `argv` will provide the runtime arguments of the `spark-submit` command, which includes `regionName`, `streamName`, `endpointUrl`, and `s3OutputPath`.

Important Note

When you integrate a real-time streaming pipeline with Spark, you can integrate either Spark Streaming or the Spark Structured Streaming API. We have integrated Spark Streaming for our use case as our data source is Kinesis Data Streams and, while writing this book, Spark Structured Streaming does not natively support Kinesis Data Streams as its source.

However, there are some open source libraries that you can integrate to take advantage of Spark Structured Streaming integration with Kinesis, or, if your data source is Kafka, then Spark Structured Streaming supports it natively.

This script should work without any modification, but you can customize it as per your ETL transformation logic.

Summary

Over the course of this chapter, we have dived deep into a real-time streaming use case, where we have integrated the data pipeline with Amazon S3, Amazon EMR, AWS Glue, and Amazon Athena.

We have covered detailed implementation steps, which you can follow to replicate the same or customize as per your use case. For our implementation, we have leveraged the Kinesis Data Generator UI tool to replicate clickstream data generation and push to Kinesis Data Streams. During your production implementation, your web application should push data to Kinesis Data Streams in real time.

At the end, we provided an overview of a few important parts of the EMR PySpark script, which can provide you with a starting point.

That concludes this chapter! Hopefully, this helped you get an idea of how you can integrate real-time streaming pipelines, and, in the next chapter, we will integrate another use case that implements UPSERT or MERGE in a data lake using the **Apache Hudi** framework.

Test your knowledge

Before moving on to the next chapter, test your knowledge with the following questions:

1. Assume that the volume of data you receive in every micro batch of the stream is very small (in KB) and, in your data lake, you plan to maintain a minimum 64-128 MB file size for better read performance. How should you design the pipeline and what trade-offs should you consider?

2. Assume, owing to infrastructure failures, that your EMR cluster got terminated but your source application is still continuously sending events to Kinesis Data Streams. When you restart your EMR cluster to resume the flow, how would you make sure that you do not lose any messages while processing the data using Spark?

Further reading

The following are a few resources you can refer to for further reading:

*   An alternative means of leveraging a Streaming application step: `https://docs.aws.amazon.com/emr/latest/ReleaseGuide/CLI_CreateStreaming.html`.

*   Monitoring a Spark Streaming application: `https://aws.amazon.com/blogs/big-data/monitor-spark-streaming-applications-on-amazon-emr/`.

11

Implementing UPSERT on S3 Data Lake with Apache Spark and Apache Hudi

In the previous two chapters, we learned how to implement a batch ETL pipeline with Amazon EMR and real-time streaming with Spark Streaming. In this chapter, we will learn how to implement **UPSERT** or merge on your Amazon S3 data lake using the **Apache Hudi** framework integrated with Apache Spark.

Amazon S3 is immutable by default, which means you cannot update the content of an object or file in S3. Instead, you have to read its content, then modify it and write a new object. Currently, as data lake and lake house architectures are becoming popular, organizations look for update capability on Amazon S3 or other object stores. Frameworks such as Apache Hudi, Apache Iceberg, and AWS Lake Formation Governed Tables have started offering ACID transactions and UPSERT capabilities on data lakes.

Apache Hudi is a popular open source framework that is integrated into Amazon EMR and AWS Glue and is also very popular within the open source community.

In this chapter, we will learn how you can integrate the Apache Hudi framework with Apache Spark to update and delete data from your data lake. To showcase this capability, we will use an EMR notebook so that you can learn how you can do interactive development on a long-running EMR cluster. The following are the topics that we will cover in this chapter:

- Apache Hudi overview
- Creating an EMR cluster and an EMR notebook
- Interactive development with Apache Spark and Apache Hudi

Getting an overview of Apache Hudi and its integration with Spark and Amazon EMR will give you a starting point to learn how to integrate the UPSERT feature in a data lake, which might help you with **General Data Protection Regulation** (**GDPR**) compliance or other regulatory requirements that compel you to update or delete data based on certain filter criteria.

Technical requirements

In this chapter, we will showcase interactive development using an EMR notebook and the Apache Spark and Apache Hudi frameworks. So, before getting started, make sure you have the following:

- An AWS Account with the ability to create Amazon S3, Amazon EMR, Amazon Athena, and AWS Glue Catalog resources
- An IAM user that can create IAM roles, which will be used to trigger or execute jobs
- Access to the Jupyter notebook that is available in our GitHub repository here: `https://github.com/PacktPublishing/Simplify-Big-Data-Analytics-with-Amazon-EMR-/tree/main/chapter_11`

Now, let's dive deep into the use case and hands-on implementation steps starting with the overview of Apache Hudi.

Check out the following video to see the Code in Action at `https://bit.ly/3svY3i9`

Apache Hudi overview

Apache Hudi is an open source framework, which is popular for providing **record-level transaction** support on top of data lakes. The Hudi framework supports integration with open file formats such as Parquet and stores additional metadata for its operations.

Apache Hudi provides several capabilities and the following are the most popular ones:

- UPSERT on top of data lakes

- Support for transactions and rollbacks

- Integration with popular distributed processing engines such as Spark, Hive, Presto, and Trino

- Automatic file compaction in data lakes

- The option to query recent update views or past transaction snapshots

Hudi supports both read and write-heavy workloads. When you write data to an Amazon S3 data lake using Hudi APIs, you have the option to specify either of the following storage types:

- **Copy on Write (CoW)**: This is the default storage type, which creates a new version of the file and stores the output in Parquet format. This is useful when you want to have the UPSERT version ready as soon as the new data is written. This is great for read-heavy workloads as you can create a new version of the file as soon as it's written, and all read workloads get the latest view.

- **Merge on Read (MoR)**: This storage type is helpful for write-heavy workloads, where the merging does not happen during the write process but instead happens on demand when a read request comes in. This stores data in a combination of Parquet and row-based Avro formats. Each new update creates a row-based incremental delta file and is compacted when needed to create a new version of the file in Parquet format.

After writing the data with the appropriate storage type, you can query it using any of the following logical views that Hudi offers:

- **Read optimized view**: This includes the latest compacted data from MoR tables and the latest committed data from CoW tables. This view does not include the delta files that are not committed or compacted yet.

- **Incremental view**: This is helpful for downstream ETL jobs as it provides an incremental change view from CoW tables.

- **Real-time view**: This view is helpful when you plan to access the latest copy of data, which merges the columnar Parquet files and the row-based Avro delta files.

The following diagram shows how CoW works for different insert and update transactions:

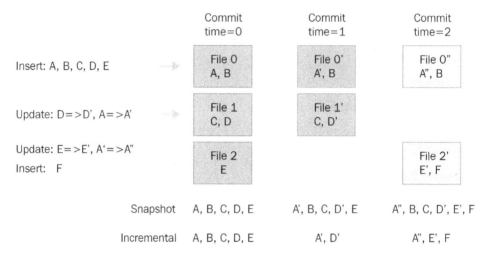

Figure 11.1 – Apache Hudi CoW commit flow
(source: https://cwiki.apache.org/confluence/display/HUDI/Design+And+Architecture)

Let's explain the preceding diagram:

- The first transaction inserts new records called **A**, **B**, **C**, **D**, and **E** that get split into three files and are marked as **Commit time=0**. At that point, both the snapshot and incremental views show the same set of records as output.

- The second transaction is an update for **D** and **A**, which creates **Commit time=1**. This will show **A**, **B**, **C**, **D**, and **E** in the snapshot and only **A** and **D** as incremental output.

- The third transaction will have an update for **A** and **E** and an insert for **F**, which creates **Commit time=2**. This will show all values, **A**, **B**, **C**, **D**, **E**, and **F**, in the snapshot and only **A**, **E**, **F** in the incremental output.

As you can see, the snapshot shows all the records as the merging happens instantly during writing action itself.

Now, let's learn how the MoR table type handles the same transactions. The following diagram represents the flow:

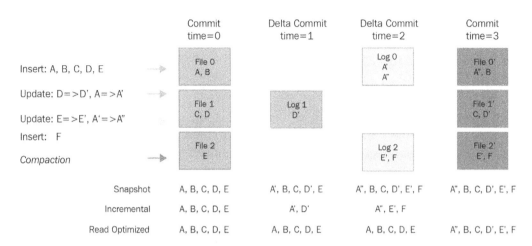

	Commit time=0	Delta Commit time=1	Delta Commit time=2	Commit time=3
Insert: A, B, C, D, E	File 0 A, B		Log 0 A' A"	File 0' A", B
Update: D=>D', A=>A'				
	File 1 C, D	Log 1 D'		File 1' C, D'
Update: E=>E', A'=>A"				
Insert: F				
	File 2 E		Log 2 E', F	File 2' E', F
Compaction				
Snapshot	A, B, C, D, E	A', B, C, D', E	A", B, C, D', E', F	A", B, C, D', E', F
Incremental	A, B, C, D, E	A', D'	A", E', F	
Read Optimized	A, B, C, D, E	A, B, C, D, E	A, B, C, D, E	A", B, C, D', E', F

Figure 11.2 – Amazon Hudi MoR commit flow

(Source: https://cwiki.apache.org/confluence/display/HUDI/Design+And+Architecture)

As explained earlier, CoW does the delta data merging instantly, whereas MoR does the merging while reading. As you can see in the preceding diagram, there is a new view called the **Read Optimized** view that shows all latest data by getting it from the MoR table and the latest committed data from CoW tables.

Now that we understand how Hudi works, let's look at some of the popular use cases for which the Hudi framework is most appropriate.

Popular use cases

The following are some of the popular use cases for which the Hudi framework is widely adopted:

- *Updating and deleting data from a data lake to meet compliance requirements*: Often for privacy regulations such as the **California Consumer Privacy Act** (**CCPA**) or the **General Data Protection Regulation** (**GDPR**), organizations are required to delete records of specific users or delete data after a specific time, which requires record-level transactions in data lakes. Hudi maintains additional metadata to keep track of the records and updates or deletes them with simple API invocations.

- *Incremental data processing*: When you are trying to set up a data lake that receives incremental **Change Data Capture** (**CDC**) data from a source system, you can use Hudi to apply the incremental changes to the data lake so that your end users can see the latest view of the data.

- *Near-real-time event streaming*: When you integrate a real-time streaming pipeline in which you might receive data for IoT systems or from a message bus such as Kinesis Data Stream, you can use Hudi with Spark Structured Streaming to update data in a data lake.

- *Having a unified data store for analytics*: When data scientists or data analysts use data for analytics, for some use cases they look for **recent real-time views** and for other use cases they look for **incremental update views**. Hudi provides the option to query different views of the data, making it a unified data store for analytics.

From the 5.28.0 release, Amazon EMR supports integration with Hudi, which means it installs Hudi-related libraries when you create an EMR cluster with Hive, Spark, or Presto.

Registering Hudi data with your Hive or Glue Data Catalog metastore

Like other Hive metastore tables, if you register your Hudi dataset with a Hive metastore then you can query your Hudi table using the Hive, Spark SQL, or Presto query engines. In addition, you can also integrate Hudi tables with your AWS Glue catalog.

As explained at the beginning of this chapter, Hudi provides two options when you write to a dataset: one is CoW and the other is MoR. When you register a table as MoR, then in your metastore, you will see two separate tables. One table has the original name that you specified, and another additional table will have a suffix of _rt to provide a real-time view of the data.

If you are using Spark to register a table with Hudi, then you should set the HIVE_SYNC_ENABLED_OPT_KEY option to true. Alternatively, you can also use the hive_sync_tool CLI utility to register your Hudi data as a metastore table in Hive or as a Glue catalog metastore.

Creating an EMR cluster and an EMR notebook

Before getting started with our use case, we need to create an EMR cluster and then create an EMR notebook that points to the EMR cluster we have created. Let's assume this EMR cluster is a long-running cluster that is active to support your development workloads as you plan to do interactive development with EMR notebooks.

Now let's learn how to create the EMR cluster and notebook.

Creating an EMR cluster

As explained in *Chapter 5*, *Setting Up and Configuring EMR Clusters*, to create an EMR cluster, follow these steps:

1. Navigate to Amazon EMR's **Create cluster** screen at `https://console.aws.amazon.com/elasticmapreduce/home?region=us-east-1#quick-create`.

2. Select **Go to advanced options** and, from the advanced options screen, select the latest stable release. We have selected the **emr-6.4.0** release because that was the latest stable release while writing this chapter. From the **Applications** list, make sure you select the **JupyterHub** and **JupyterEnterpriseGateway** applications with **Spark** as they will be needed for the EMR notebook.

The following is a screenshot of the EMR release and applications we have selected:

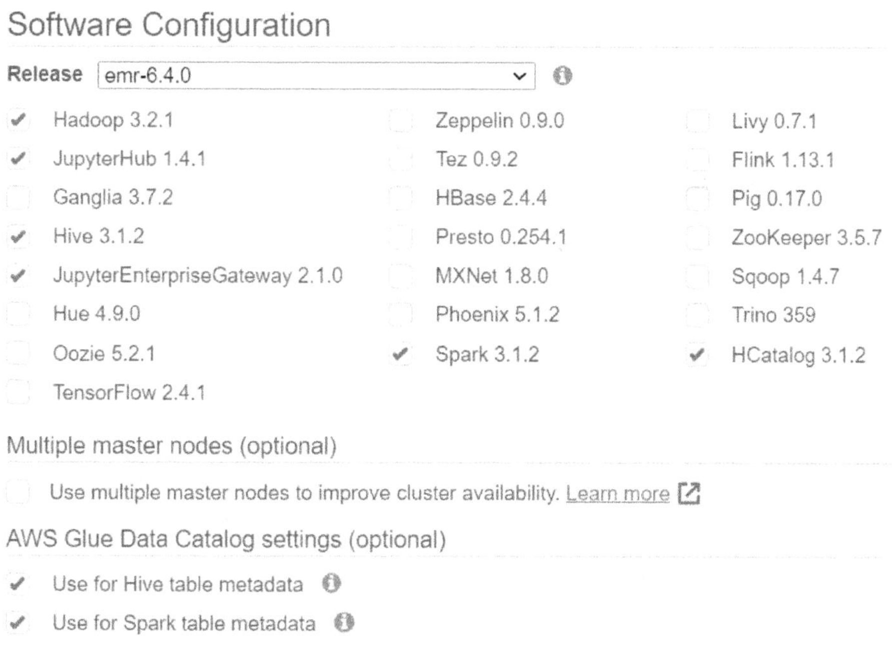

Software Configuration

Release emr-6.4.0

✔ Hadoop 3.2.1	Zeppelin 0.9.0	Livy 0.7.1
✔ JupyterHub 1.4.1	Tez 0.9.2	Flink 1.13.1
Ganglia 3.7.2	HBase 2.4.4	Pig 0.17.0
✔ Hive 3.1.2	Presto 0.254.1	ZooKeeper 3.5.7
✔ JupyterEnterpriseGateway 2.1.0	MXNet 1.8.0	Sqoop 1.4.7
Hue 4.9.0	Phoenix 5.1.2	Trino 359
Oozie 5.2.1	✔ Spark 3.1.2	✔ HCatalog 3.1.2
TensorFlow 2.4.1		

Multiple master nodes (optional)

Use multiple master nodes to improve cluster availability. Learn more 🗗

AWS Glue Data Catalog settings (optional)

✔ Use for Hive table metadata ℹ

✔ Use for Spark table metadata ℹ

Figure 11.3 – Amazon EMR Create cluster Software Configuration screen

3. Next, under **Hardware configurations**, enable **Cluster scaling** with the default values for **EMR-managed scaling**. Then, under **Auto-termination**, disable the **Enable auto-termination** checkbox.

4. In **General Cluster Settings**, keep everything as the default values and then, on the **Security and access** screen, select **EC2 key pair**, with which you can SSH to the EMR cluster's master node. The following screenshot shows the values we have selected:

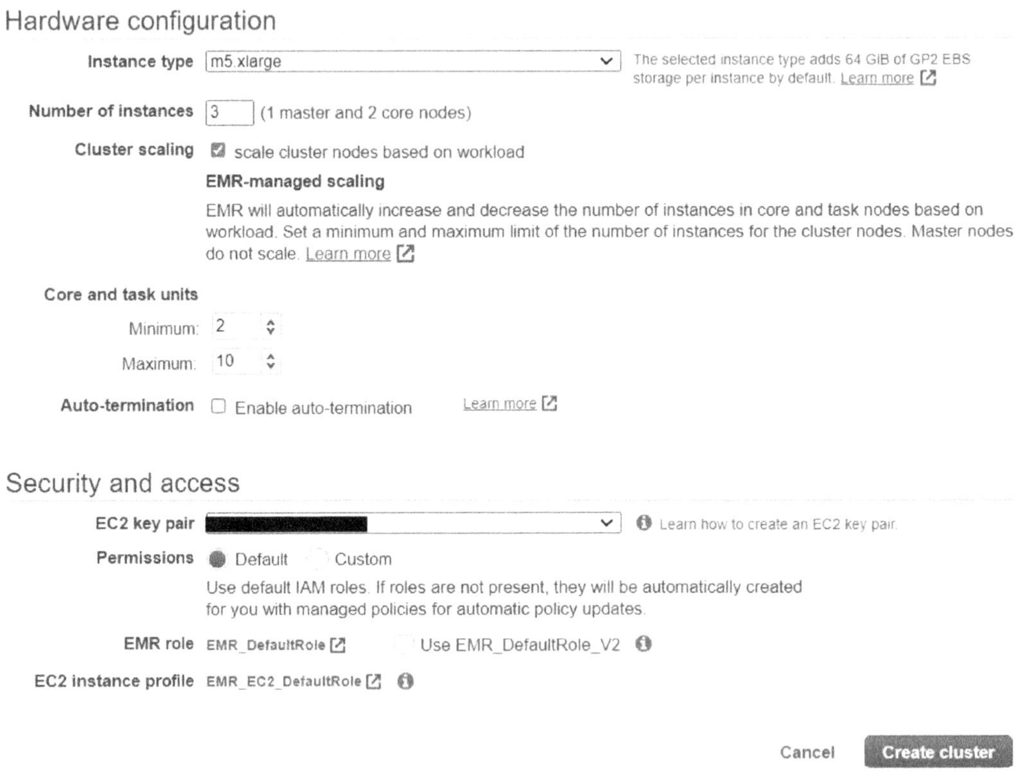

Hardware configuration

Instance type   `m5.xlarge`   The selected instance type adds 64 GiB of GP2 EBS storage per instance by default. Learn more

Number of instances   `3`   (1 master and 2 core nodes)

Cluster scaling   ☑ scale cluster nodes based on workload

EMR-managed scaling

EMR will automatically increase and decrease the number of instances in core and task nodes based on workload. Set a minimum and maximum limit of the number of instances for the cluster nodes. Master nodes do not scale. Learn more

Core and task units

Minimum: 2

Maximum: 10

Auto-termination   ☐ Enable auto-termination   Learn more

Security and access

EC2 key pair   ▮▮▮▮▮▮▮   ❶ Learn how to create an EC2 key pair.

Permissions   ● Default   ○ Custom

Use default IAM roles. If roles are not present, they will be automatically created for you with managed policies for automatic policy updates.

EMR role   EMR_DefaultRole   Use EMR_DefaultRole_V2 ❶

EC2 instance profile   EMR_EC2_DefaultRole ❶

Cancel   **Create cluster**

Figure 11.4 – Amazon EMR Create cluster screen – Hardware and security configuration

5. Then select **Create cluster**, which will take you to the EMR cluster detail screen and have a status of **Starting**.

After a few minutes, you will notice the cluster status changes to **Running** when the initial **Setup hadoop debugging** default job runs, and after the job is complete, it changes to **Waiting,** which means all the resources are provisioned and we are ready to submit jobs to the cluster.

Let's see how to create an EMR notebook that points to our EMR cluster.

Creating an EMR notebook

To create an EMR notebook, follow these steps:

1. Navigate to the EMR notebook list in the EMR console at `https://console.
    aws.amazon.com/elasticmapreduce/home?region=us-east-
    1#notebooks-list`.

2. Click **Create notebook**, which will open a form on the web interface to configure
    your notebook.

3. On the **Create notebook** form, add a notebook name and then, in the **Clusters**
    field, select the **Choose an existing cluster** option and then click **Choose**, which
    will open a pop-up overlay with a list of the EMR clusters you have.

4. Select the EMR cluster you created in the previous step and click **Choose cluster**.

5. Keep the rest of the field values as the defaults and then click **Create notebook**.

The following screenshot shows the **Create notebook** form in the EMR console:

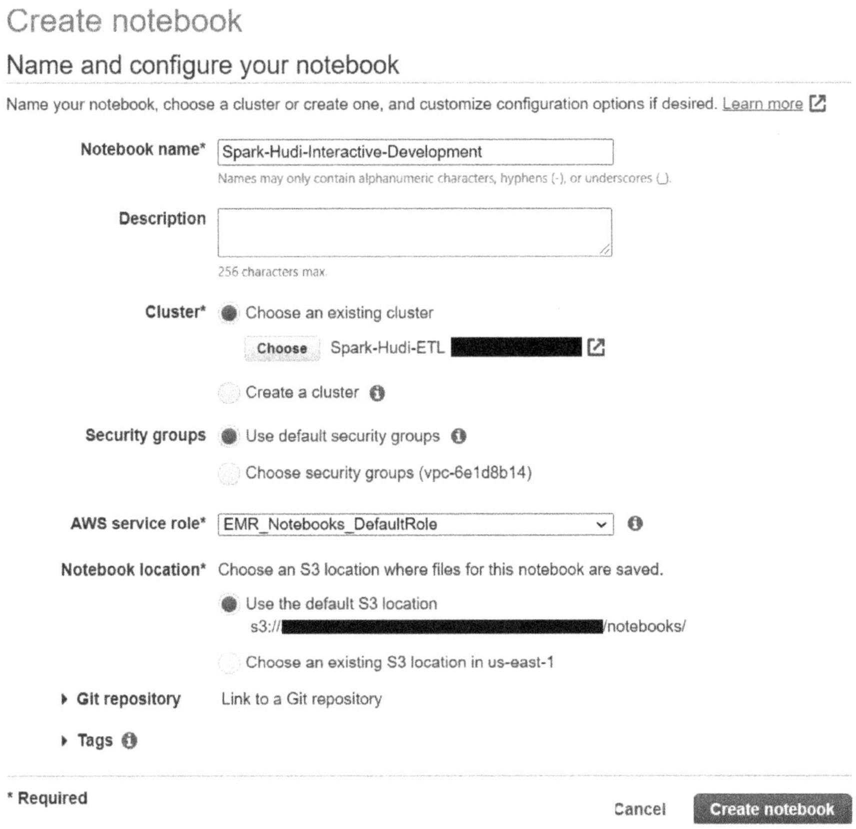

Figure 11.5 – The Create notebook form

Clicking on the **Create notebook** button will take you to the EMR Notebook's detail page, with a status of **Starting**. In a few minutes, the status will change to **Ready**.

Now you can click the **Open in Jupyter** button on the notebook detail page, as shown in the following screenshot:

Figure 11.6 –EMR notebook detail page

This will take you to Jupyter Notebook, where you can create interactive notebooks for development. Jupyter Notebook provides options to create a notebook in languages such as PySpark, SparkR, Python 3, or the Linux command-line Terminal.

As a next step, before moving onto the Spark and Hudi implementation, let's first create an Amazon S3 bucket, which we can use to store Hudi datasets.

Creating an Amazon S3 bucket

Please refer to the following steps to create the S3 bucket, which we have followed in previous chapters too:

1. Navigate to Amazon S3 console at `https://s3.console.aws.amazon.com/s3/home?region=us-east-1#`.

2. From the buckets list, choose **Create Bucket**. This will open a form on the web interface to provide your bucket name and related configurations.

   We have called our bucket `hudi-data-repository` and kept everything else as the default values.

3. Then, click the **Create bucket** button to create the bucket.

The following screenshot shows the AWS console, using which we have created the bucket:

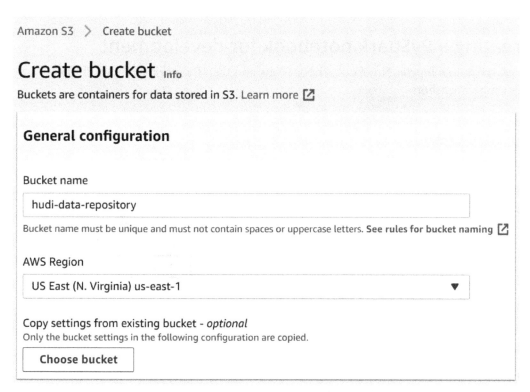

Figure 11.7 – Amazon S3 Create bucket page

As we have now created all the resources, next we will dive deep into our use case implementation with Spark and Hudi.

Interactive development with Spark and Hudi

Our EMR cluster and notebook are now ready for use. Let's learn how to do interactive development using an EMR notebook.

For interactive development, we are considering a use case where we will integrate the Hudi framework with Spark to do UPSERT (update/merge) operations on top of an S3 data lake.

Let's navigate to our EMR notebook to get started.

Creating a PySpark notebook for development

To get started, in Jupyter Notebook, choose **New** and then **PySpark**, as shown in the following screenshot:

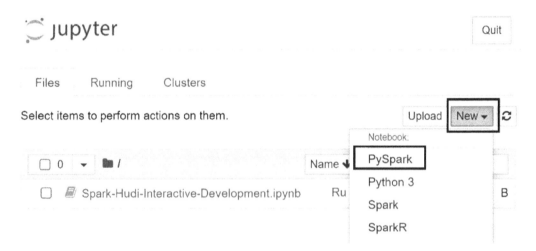

Figure 11.8 – The Jupyter Notebook landing page

This will create a new PySpark notebook. In every cell, you can write scripts and execute them line by line for easy development or debugging.

Next, we will learn how to integrate Hudi libraries with the notebook.

Integrating Hudi with our PySpark notebook

By default, Hudi libraries are not available in our EMR notebook. To make them available, you need to copy the Hudi **Java ARchive (JAR)** files from the EMR cluster's master node to HDFS so that the EMR notebook can refer to them. Follow these steps to do so:

1. Navigate to your EMR cluster list at `https://console.aws.amazon.com/elasticmapreduce/home?region=us-east-1#cluster-list`.

2. Choose the cluster you have created, which will be in the **Waiting** state. This will take you to the cluster detail page.

3. Copy the cluster's **Master public DNS** URL and SSH to the master node using PuTTY or an equivalent tool. You can also click the **Connect to the Master Node Using SSH** link below the DNS URL to learn how you can SSH to the master node using Windows, macOS, and Linux systems.

The following screenshot shows the cluster details page, from where you can copy the master node's public DNS URL:

Figure 11.9 – EMR cluster details page

For our implementation, we used PuTTY to SSH to the EMR master node, and to do that, we specified the master public DNS URL in the **Session | Host Name** field and then specified the EC2 key pair private key (.ppk) under **Connection | SSH | Auth**, as shown in the following screenshot:

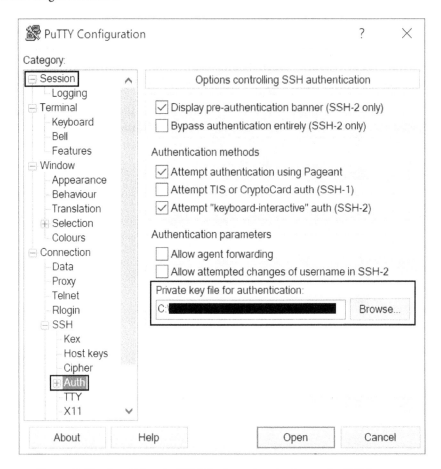

Figure 11.10 – Specifying the PPK file for authentication to the EC2 instance

When you click **Open** and try to connect for the first time, PuTTY will ask you to confirm whether you trust the connection. Click **Yes**, as shown in the following screenshot:

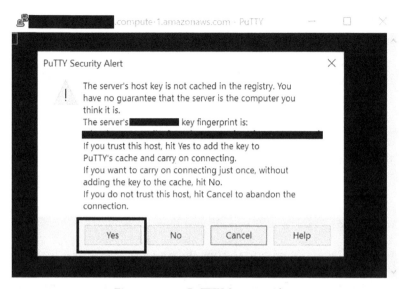

Figure 11.11 – PuTTY Security Alert

After you click **Yes**, you will be prompted to enter the login user name, for which you should type hadoop. That will connect successfully, as shown in the following screenshot:

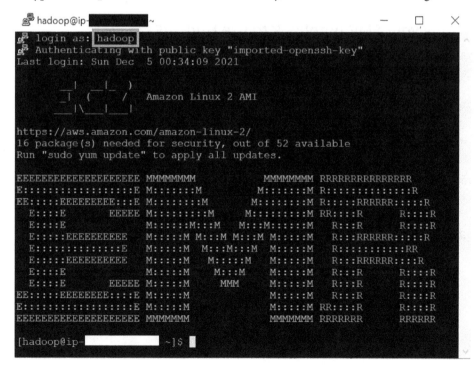

Figure 11.12 – PuTTY SSH login

Once you are successfully logged in, you can execute the following commands to copy the Hudi JAR files from the master node's local filesystem to HDFS.

First, create a new directory path in HDFS. We have used the `/applications/hudi/lib` path in HDFS, but you can use your own path. You need to keep a note of this path so that you can use it in your EMR notebook:

```
hdfs dfs -mkdir -p /applications/hudi/lib
```

Then execute the following command to copy the `hudi-spark-bundle.jar` file to HDFS:

```
hdfs dfs -copyFromLocal /usr/lib/hudi/hudi-spark-bundle.jar /
applications/hudi/lib/hudi-spark-bundle.jar
```

Finally, execute the following command to copy the `spark-avro.jar` file to HDFS:

```
hdfs dfs -copyFromLocal /usr/lib/spark/external/lib/spark-avro.
jar /applications/hudi/lib/spark-avro.jar
```

Once these two files are copied to HDFS, you can refer to them in your EMR notebook.

Configuring your EMR notebook to use Hudi JARs

After you have the Hudi JARs available in HDFS, you can navigate back to your EMR PySpark notebook and execute the following command:

```
%%configure -f
{
    "conf":  {
                "spark.jars":"hdfs:///applications/hudi/lib/hudi-
spark-bundle.jar,hdfs:///applications/hudi/lib/spark-avro.jar",
                "spark.sql.hive.convertMetastoreParquet":"false",
                "spark.serializer":"org.apache.spark.serializer.
KryoSerializer"
            }
}
```

This makes the Hudi JARs available to SparkContext so that the Spark code can use its libraries. The following screenshot shows the output you should see in your notebook:

```
In [1]: %%configure -f
        {
            "conf":  {
                    "spark.jars":"hdfs:///applications/hudi/lib/hudi-spark-bundle.jar,hdfs:///applications/hu
                    "spark.sql.hive.convertMetastoreParquet":"false",
                    "spark.serializer":"org.apache.spark.serializer.KryoSerializer"
                }
        }

Current session configs: {'conf': {'spark.jars': 'hdfs:///applications/hudi/lib/hudi-spark-
bundle.jar,hdfs:///applications/hudi/lib/spark-avro.jar', 'spark.sql.hive.convertMetastoreParquet':
'false', 'spark.serializer': 'org.apache.spark.serializer.KryoSerializer'}, 'proxyUser': 'assumed-
role_developer_missakti-Isengard', 'kind': 'pyspark'}

No active sessions.
```

Figure 11.13 – Additional configuration settings

After our setup is complete, let's look at a few Hudi APIs and example scripts using which you can create CoW or MoR tables and can do transactions on top of the tables.

Executing Spark and Hudi scripts in your notebook

Now that our notebook is ready, with all the required JARs, let's dive into Hudi APIs that enable ACID transactions on top of our data lake. We will use PySpark scripts to interact with Hudi functions.

For our use case, we have created an example product inventory dataset, which has `product_id`, `category`, `product_name`, `quantity_available`, and `last_update_time` fields. We will write the input data to our S3 data lake in Hudi format, then will update and delete some records and then query to validate the transactional updates.

Let's now work throughout this step by step.

Inserting new product inventory data into our S3 data lake

The first step for us is to ingest new product inventory data in our data lake. `product_id` is the unique key to identify a product and the `category` field is used to partition the data in S3.

With the following code, we create a Spark DataFrame using some product records:

```
# Create a DataFrame that represents Product Inventory
inputDF = spark.createDataFrame(
    [
        ("100", "Furniture", "Product 1", "25",
"2021-12-01T09:51:39.340396Z"),
        ("101", "Cosmetic", "Product 2", "20",
"2021-12-01T10:14:58.597216Z"),
        ("102", "Furniture", "Product 3", "30",
"2021-12-01T11:51:40.417052Z"),
        ("103", "Electronics", "Product 4", "10",
"2021-12-01T11:51:40.519832Z"),
        ("104", "Electronics", "Product 5", "50",
"2021-12-01T11:58:00.512679Z")
    ],
    ["product_id", "category", "product_name", "quantity_
available", "last_update_time"]
)
```

Next, we create the hudiOptions configuration variable, which specifies the Hudi parameters that represent the table name, record key, partition key, and more:

```
# Specify common DataSourceWriteOptions in the single
hudiOptions variable
hudiOptions = {
'hoodie.table.name': 'product_inventory',
'hoodie.datasource.write.recordkey.field': 'product_id',
'hoodie.datasource.write.partitionpath.field': 'category',
'hoodie.datasource.write.precombine.field': 'last_update_time',
'hoodie.datasource.hive_sync.enable': 'true',
'hoodie.datasource.hive_sync.table': 'product_inventory',
'hoodie.datasource.hive_sync.partition_fields': 'category',
'hoodie.datasource.hive_sync.partition_extractor_class': 'org.
apache.hudi.hive.MultiPartKeysValueExtractor'
}
```

Next, we can write the product records to Amazon S3 using the `hudiOptions` configuration variable:

```
# Write the product Inventory DataFrame as a Hudi dataset to S3
inputDF.write.format('org.apache.hudi') \
.option('hoodie.datasource.write.operation', 'insert') \
.options(**hudiOptions) \
.mode('overwrite') \
.save('s3://hudi-data-repository/product-inventory/')
```

As you can see, the `hoodie.datasource.write.operation` parameter has a value of `insert` to represent its new insert. The following screenshot shows the output of the execution:

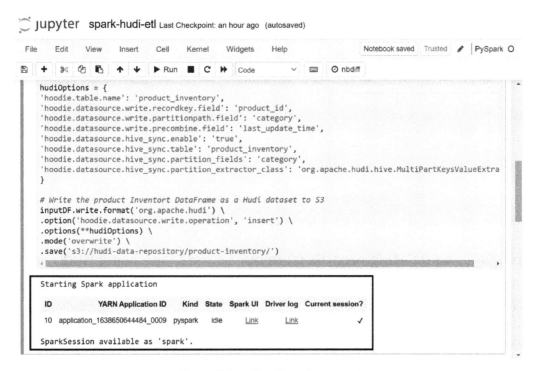

Figure 11.14 – Data ingestion output

After the successful execution of the preceding code, you can navigate to your S3 bucket to validate the data being written in Hudi format with partition folders for the `category` field. The following screenshot shows the output:

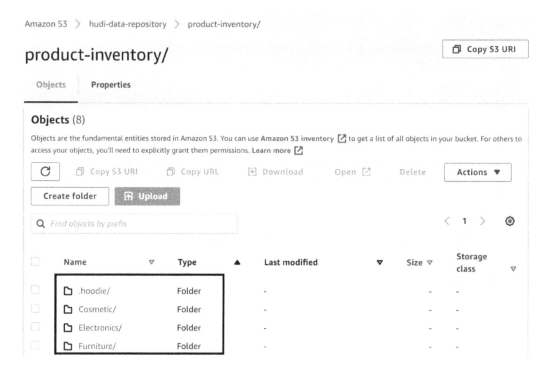

Figure 11.15 – S3 output with partitioning folders

After the initial data is written, we can try to update and delete a few records.

Updating and deleting records from our S3 data lake using Hudi and Spark

Before updating records, let's first query the data and validate how it looks. The following code block reads data from `s3://hudi-data-repository/product-inventory` with `/*/*` so that it reads all partitions:

```
# Read the Hudi dataset from S3 and validate your field output
HudiProductDF = spark.read.format('org.apache.hudi').
load('s3://hudi-data-repository/product-inventory' + '/*/*')
HudiProductDF.select("product_id", "category", "product_name",
"quantity_available", "last_update_time").show()
```

After executing this code, you should see the following output, which shows the data of all the five products you have ingested:

```
In [3]:  # Read the Hudi dataset from S3 and validate your field output
         HudiProductDF = spark.read.format('org.apache.hudi').load('s3://hudi-data-repository/product-inventory
         HudiProductDF.select("product_id", "category", "product_name", "quantity_available", "last_update_time
         ◀
```

▸ Spark Job Progress

```
+----------+-----------+------------+------------------+-------------------+
|product_id|   category|product_name|quantity_available|   last_update_time|
+----------+-----------+------------+------------------+-------------------+
|       101|   Cosmetic|   Product 2|                20|2021-12-01T10:14:...|
|       103|Electronics|   Product 4|                10|2021-12-01T11:51:...|
|       104|Electronics|   Product 5|                50|2021-12-01T11:58:...|
|       102|   Furniture|   Product 3|                30|2021-12-01T11:51:...|
|       100|   Furniture|   Product 1|                25|2021-12-01T09:51:...|
+----------+-----------+------------+------------------+-------------------+
```

Figure 11.16 – Data ingestion output

Next, let's assume you have sold one unit of product ID 102 and would like to update its quantity to 29. The following code shows how to do this. As you can see, the hoodie. datasource.write.operation parameter value is set to upsert:

```
# Update quanity of product_id 102
from pyspark.sql.functions import col,lit
newDF = inputDF.filter(inputDF.product_id==102).
withColumn('quantity_available',lit('29'))
newDF.write \
.format('org.apache.hudi') \
.option('hoodie.datasource.write.operation', 'upsert') \
.options(**hudiOptions) \
.mode('append') \
.save('s3://hudi-data-repository/product-inventory/')
```

Next, let's assume you have stopped selling product ID 101 and would like to delete it from your inventory table. The following code block shows how to delete a product using Hudi. As you can see, we have passed an additional parameter, hoodie.datasource. write.payload.class, with a value of org.apache.hudi.common.model. EmptyHoodieRecordPayload to represent deleting the record:

```
# Delete product record with ID 101
deleteDF = inputDF.filter(inputDF.product_id==101)
deleteDF.write \
```

```
.format('org.apache.hudi') \
.option('hoodie.datasource.write.operation', 'upsert') \
.option('hoodie.datasource.write.payload.class', 'org.apache.
hudi.common.model.EmptyHoodieRecordPayload') \
.options(**hudiOptions) \
.mode('append') \
.save('s3://hudi-data-repository/product-inventory/')
```

After the executing both the update and delete transactions, we can validate the output by executing the following PySpark script, which reads the updated data from S3:

```
# Read from S3 to validate the update and delete record
HudiProductNewDF = spark.read.format('org.apache.hudi').
load('s3://hudi-data-repository/product-inventory' + '/*/*')
HudiProductNewDF.select("product_id", "category", "product_
name", "quantity_available", "last_update_time") \
.orderBy("product_id").show()
```

The following screenshot shows the output of the execution. The record with `product_id` 101 is missing and the record with `product_id` 102 has a quantity value of 29:

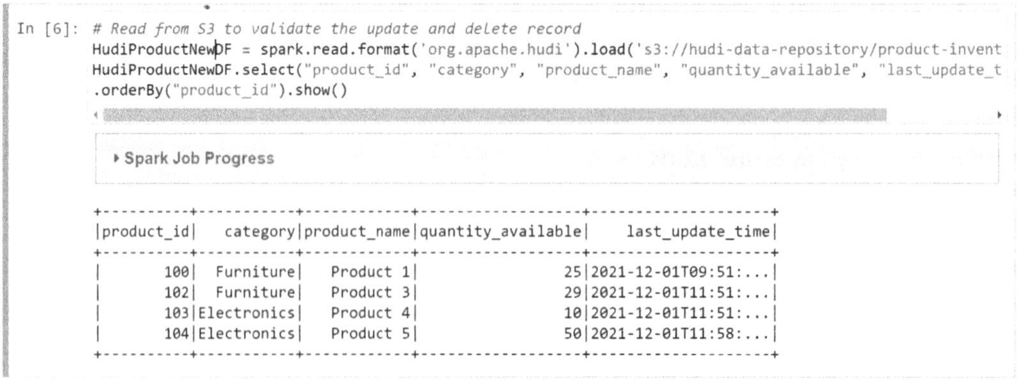

Figure 11.17 – Output after update and delete

So far, we have learned how to create Hudi tables, ingest new data, and update and delete datasets using Spark and Hudi. Next, let's learn how to query incremental data and some of the additional metadata attributes Hudi provides.

Querying incremental data using Hudi

In our previous execution, when we tried to query data, we defined specific columns to validate our data. Now let's try to print the complete dataset and make a note of the additional attributes Hudi appends:

```
# List all columns on the dataframe to showcase additional
metadata fields Hudi appends
HudiProductNewDF.show()
```

The following screenshot highlights the additional attributes Hudi appends, and all of them are prefixed with _:

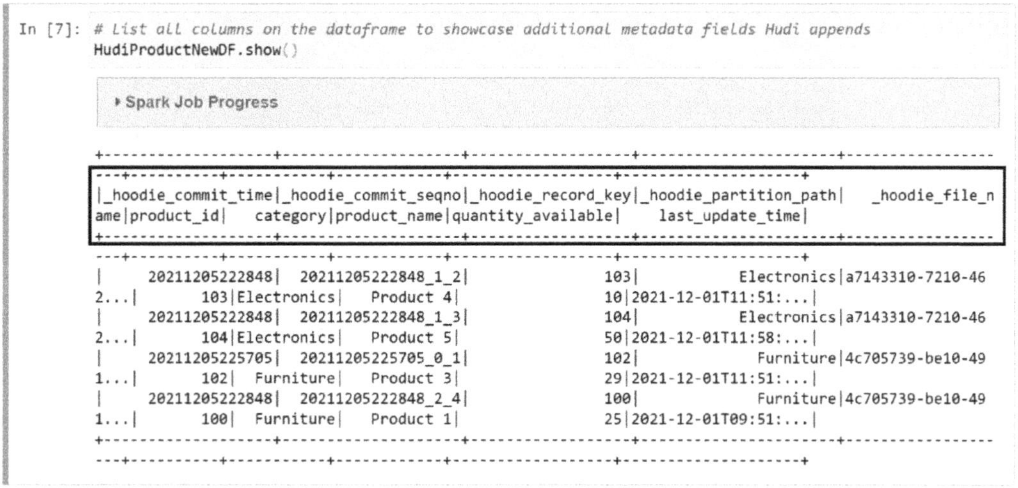

Figure 11.18 – Hudi attributes

As you can see in this output, all the records have a hoodie commit time of 20211205222848, except product_id 102, which has a hoodie commit time of 20211205225705 because we updated that record after our initial data ingestion.

Next, let's try to query the incremental data by listing all the records changed after a certain timestamp, such as `20211205222848`. The following code block shows how to query incremental data using Hudi:

```
# Incremental query output, that fetches change data beyond
certain time
incrementalQueryOptions = {
  'hoodie.datasource.query.type': 'incremental',
  'hoodie.datasource.read.begin.instanttime': "20211205222848",
}
incQueryDF = spark.read.format('org.apache.hudi').
options(**incrementalQueryOptions) \
.load('s3://hudi-data-repository/product-inventory')
incQueryDF.show()
```

The following screenshot highlights the output we get after executing the incremental query and, as you can see, there is one record, that is, `product_id 102`:

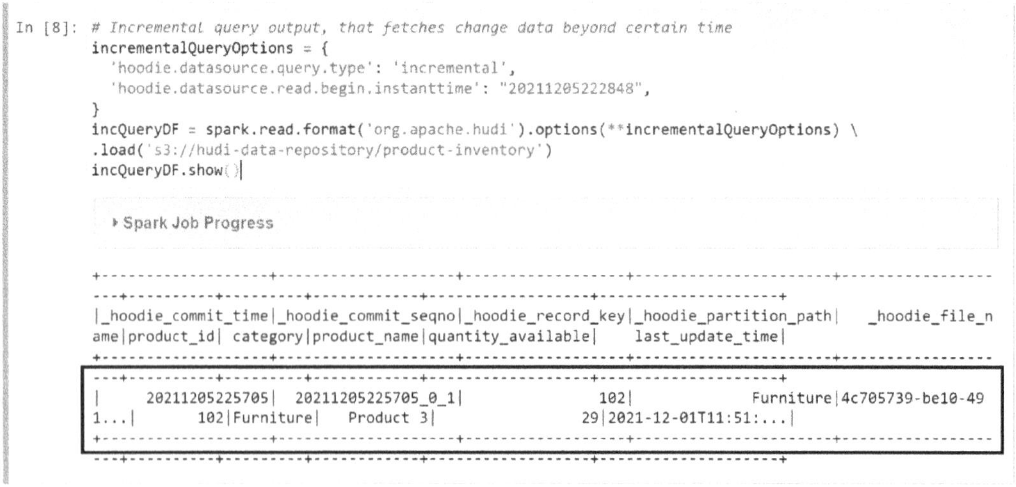

```
In [8]: # Incremental query output, that fetches change data beyond certain time
        incrementalQueryOptions = {
          'hoodie.datasource.query.type': 'incremental',
          'hoodie.datasource.read.begin.instanttime': "20211205222848",
        }
        incQueryDF = spark.read.format('org.apache.hudi').options(**incrementalQueryOptions) \
        .load('s3://hudi-data-repository/product-inventory')
        incQueryDF.show()

        ▸ Spark Job Progress

+------------------+-------------------+------------------+----------------------+-----------------
---+----------+---------+------------+------------------+--------------------+
|_hoodie_commit_time|_hoodie_commit_seqno|_hoodie_record_key|_hoodie_partition_path|   _hoodie_file_n
ame|product_id| category|product_name|quantity_available|    last_update_time|
+------------------+-------------------+------------------+----------------------+-----------------
---+----------+---------+------------+------------------+--------------------+
|     20211205225705|  20211205225705_0_1|               102|             Furniture|4c705739-be10-49
1...|       102|Furniture|   Product 3|                29|2021-12-01T11:51:...|
+------------------+-------------------+------------------+----------------------+-----------------
---+----------+---------+------------+------------------+--------------------+
```

Figure 11.19 – Hudi incremental query output

In this section, we have walked you through the Spark and Hudi code using which you can insert, update, delete, and query Hudi datasets. Having the UPSERT capability on top of our S3 data lake provides a lot of flexibility in terms of meeting compliance requirements or having a golden copy for querying. This should give you a starting point for your Hudi implementation, and you can refer to the Hudi documentation for additional information.

> **Note**
>
> During this chapter, we have executed Hudi scripts using an EMR notebook, but you can automate the execution by saving the script into a Python file and submitting it as an EMR step job, as explained in *Chapter 9, Implementing Batch ETL Pipeline with Amazon EMR and Apache Spark*.

Summary

Over the course of this chapter, we have dived deep into Apache Hudi and looked at its features, use cases, and how it is integrated with AWS and Amazon EMR.

We have covered how to create an EMR notebook that points to a long-running EMR cluster and how to use the notebook for interactive development. To showcase interactive development, we explained a small use case using Spark and Hudi, which can enable you to do UPSERT transactions on top of a data lake.

That concludes this chapter! Hopefully, this has helped you get an idea of how to use EMR notebooks for interactive development. In the next chapter, we will explain how to build a workflow to build a data pipeline using Amazon EMR.

Test your knowledge

Before moving on to the next chapter, test your knowledge with the following questions:

1. Assume your data science team is using EMR notebooks for their interactive development and they are primarily using Python 3 for machine learning model development. When they started executing the Python code, they found some of their scripts are not getting executed; they get an error stating that the Python module does not exist. How would you make the additional Python modules available in the EMR notebook so that your data science team can continue executing their scripts for machine learning model development?

2. Assume you have an S3 data lake on top of which you have created Hudi tables for ACID transactions and UPSERT. You are updating records as they change, which creates multiple versions of the records in the Hudi table. You have received a business requirement to find the value of a specific column at a specific time. How would you fulfill that requirement using Hudi libraries?

Further reading

Here are a few resources you can refer to for further reading:

- Apache Hudi documentation: `https://hudi.apache.org/`

- EMR and Hudi integration: `https://docs.aws.amazon.com/emr/latest/ReleaseGuide/emr-hudi.html`

- Considerations and limitations while using Hudi with Amazon EMR: `https://docs.aws.amazon.com/emr/latest/ReleaseGuide/emr-hudi-considerations.html`

- Learn more about EMR Notebooks: `https://docs.aws.amazon.com/emr/latest/ManagementGuide/emr-managed-notebooks.html`

12

Orchestrating Amazon EMR Jobs with AWS Step Functions and Apache Airflow/MWAA

In the previous few chapters, we explained how you can leverage the EMR cluster for on-demand ETL jobs or long-running clusters that either execute a real-time streaming application or serve as a backend for interactive development using notebooks. But when we build a data pipeline to automate data ingestion, cleansing, or transformations, we look for orchestration tools with which we can build workflows that either get kicked off through a schedule or through an event.

There are two primary orchestration tools – **AWS Step Functions** and **Apache Airflow**, which are very popular in building data pipelines with **Amazon EMR**. AWS also provides a managed offering of Airflow, called **Amazon Managed Workflows for Apache Airflow** (**MWAA**).

In this chapter, we will provide an overview of AWS Step Functions and MWAA services and then explain how you can leverage them to orchestrate a data pipeline that can create EMR clusters, submit jobs, and terminate clusters as required. The following are the high-level topics we will be covering:

- Overview of AWS Step Functions
- Integrating AWS Step Functions to orchestrate EMR jobs
- Overview of Apache Airflow and MWAA
- Integrating Airflow to trigger EMR jobs

Getting an overview of these orchestration tool options will give you a starting point and you should be able to build more complex data pipelines that not only integrate Amazon EMR jobs but also other AWS and non-AWS services to automate your workflow.

Technical requirements

In this chapter, we will showcase the features of AWS Step Functions and MWAA and demonstrate how you can integrate them to trigger EMR jobs. So, before getting started, make sure you have the following requirements to hand.

- An AWS account with access to create Amazon S3, Amazon EMR, AWS Step Functions, and MWAA resources
- An IAM user with access to create IAM roles, which will be used to trigger or execute jobs

Now let's get an overview of these orchestration tools and learn how we can integrate them.

Overview of AWS Step Functions

AWS Step Functions is a serverless workflow service that provides integration with several AWS services *natively*, which means you can create a workflow that is able to integrate or invoke actions of all the supported AWS services.

AWS Step Functions provides both a visual interface and a JSON based-definition approach to design workflows. With the visual interface, you can drag and drop different AWS service actions and modify their parameters to integrate a workflow. In addition to the visual interface, Step Functions also provides the option to code your workflow with a JSON-based definition called a **state machine**, where each step is referred to as a **state**. Step Functions also provides a few sample projects that are frequently in use, which you can inherit and modify for your use case.

You can integrate AWS Step Functions to automate IT business processes or build data or machine learning pipelines, or can integrate it to design real-time, event-based streaming applications. When you start designing a workflow using Step Functions, you can choose either of the following types.

- **Standard:** This is great for long-running batch workflows that can be related to building data analytics pipelines or automating IT processes that need durable and auditable workflows. Standard state machines can stay active for a year and are billed based on the number of state transitions.

- **Express**: This is useful when you need to build an event-driven workflow that has a higher volume of requests compared to a batch workload that runs on a schedule. These workflows have a maximum execution timeout of 5 minutes and are billed based on the number of requests and the duration of the workflow.

When you design a workflow using Step Functions, each step or state might be any of the following types:

- `Pass`: This allows you to skip and move to the next step.

- `Task`: This allows you to invoke the actions of other AWS services, such as invoking an AWS Lambda function or triggering an EMR Spark Job.

- `Choice`: This is similar to **Switch-Case** statements in programming languages, where you can define conditions and actions based on parameter values.

- `Wait`: This enables you to introduce a wait time (in seconds) into the workflow.

- `Succeed`: This enables you to successfully terminate the workflow.

- `Fail`: This enables you to terminate the workflow with a fail status.

- `Parallel`: This is an important feature that enables you to run multiple tasks in parallel.

- `Map`: This is also an important type that allows you to iterate the complete workflow steps for a given set of records.

Out of all the state types that AWS Step Functions supports, the `Task` type is the most commonly used one as that enables you to invoke actions of other AWS services to build the workflow. Now, let's learn how you can leverage AWS Step Functions to build a workflow using Amazon EMR actions.

Integrating AWS Step Functions to orchestrate EMR jobs

AWS Step Functions supports `createCluster`, `createCluster.sync`, `terminateCluster`, `terminateCluster.sync`, `addStep`, `cancelStep`, `setClusterTerminationProtection`, `modifyInstanceFleetByName`, and `modifyInstanceGroupByName` EMR actions, which provides a great flexibility to build workflows on top of EMR.

Let's assume that you would like to build a workflow that gets triggered as soon as a file arrives in S3 and the objective of the workflow is to execute a Spark + Hudi job to process the input file. The workflow is supposed to create a transient EMR cluster, submit a Spark job that does ETL transforms, and then, upon completion of the job, terminate the cluster. You can easily build this workflow using AWS Step Functions' `createCluster`, `addStep`, and `terminateCluster` actions.

The following JSON definition is an example of a Step Functions' step that is of the `Task` type and invokes the EMR `createCluster` action with parameters that are required to create the cluster:

```
"Launch_EMR_Cluster":{
  "Type":"Task",
"Resource":"arn:aws:states:::elasticmapreduce:createCluster.
sync",
  "Parameters":{
     "Name":"StepFn-EMR-Hudi",
     "ServiceRole":"EMR_DefaultRole",
     "JobFlowRole":"EMR_EC2_DefaultRole",
     "EbsRootVolumeSize":10,
     "ReleaseLabel":"emr-6.4.0",
     "Applications":[{"Name":"Hadoop"},{"Name":"Spark"},{"Name":
"Hive"},{"Name":"Livy"}],
     "LogUri":"s3://<bucket-name>/emr/logs",
```

```
    "ManagedScalingPolicy":{
        "ComputeLimits":{
            "MaximumCapacityUnits":2,
            "MaximumCoreCapacityUnits":2,
            "MaximumOnDemandCapacityUnits":2,
            "MinimumCapacityUnits":1,
            "UnitType":"InstanceFleetUnits"
        }
    },
    "VisibleToAllUsers":true,
    "Instances":{
        "KeepJobFlowAliveWhenNoSteps":true,
        "Ec2KeyName":"<key-pair-name>",
        "Ec2SubnetId":"<subnet-id>",
        "InstanceFleets":[
            {
                "InstanceFleetType":"MASTER",
                "Name":"Master",
                "TargetOnDemandCapacity":1,
                "InstanceTypeConfigs":[{"InstanceType":"m5.
xlarge"}]
            },
            {
                "InstanceFleetType":"CORE",
                "TargetOnDemandCapacity":1,
        "InstanceTypeConfigs":[{"InstanceType":"m5.xlarge"}] ,
            }
        ]
    }
},
"Next":"Copy_Hudi_JARs"
```

There are a few variables, including <bucket-name>, <key-pair-name>, and
<subnet-id>, that you must replace before integrating this state.

The following JSON definition represents an example of a Step Functions' step that adds a step or job to the EMR cluster. You may observe that it references the `$.ClusterId` variable from the previous `createCluster` action to submit a job to the same cluster:

```
"Trigger_Spark_Job":{
  "Type":"Task",
  "ResultPath":"$.Result",
  "Catch":[
      {
          "ErrorEquals":["States.ALL"],
          "ResultPath":"$.error-info",
          "Next":"Terminate_EMR_Cluster"
      }
  ],
  "Resource":"arn:aws:states:::elasticmapreduce:addStep.sync",
  "Parameters":{
      "ClusterId.$":"$.ClusterId",
      "Step":{
          "Name":"Spark Transform Step",
          "ActionOnFailure":"CONTINUE",
          "HadoopJarStep":{
              "Jar":"command-runner.jar",
              "Args":["spark-submit", "--deploy-mode", "cluster",
"--jars", "/usr/lib/hudi/hudi-spark-bundle.jar", "--conf",
"spark.serializer=org.apache.spark.serializer.KryoSerializer",
"s3://<bucket-name>/<script-path-name>.py"
              ]
          }
      }
  },
  "Next":"Terminate_EMR_Cluster"
},
```

As you can see, it receives the PySpark script path as a parameter. You must replace the `<bucket-name>` and `<script-path-name>` variables before integrating this into your parent state machine definition.

After you complete all your ETL transformation steps, if you plan to integrate the
terminateCluster step, then you can refer to the following JSON definition that
invokes the EMR terminateCluster step:

```
"Terminate_EMR_Cluster":{
  "Type":"Task",
  "Resource":"arn:aws:states:::elasticmapreduce:terminateCluster.
sync",
  "Parameters":{"ClusterId.$":"$.ClusterId"},
  "End":true
}
```

Please note that .sync in all these EMR actions represents the fact that Step Functions
will wait for the step to be completed before moving on to the next step.

The complete state machine definition is available here: https://github.com/
PacktPublishing/Simplify-Big-Data-Analytics-with-Amazon-EMR-/
blob/main/chapter_12/emr-cluster-job-step-functions.json. This
can be downloaded, modified, and integrated into your AWS account. The following
is a snapshot of Step Functions' visual representation of the workflow when it is being
executed. If you notice, it has an additional step after creating a cluster, which is to copy
the Hudi JARs. If you recollect from *Chapter 11, Implementing UPSERT on S3 Data Lake
with Apache Hudi*, we have performed this manual JAR copy operation by SSHing to
master nodes, which can also be automated using the Step Functions' task.

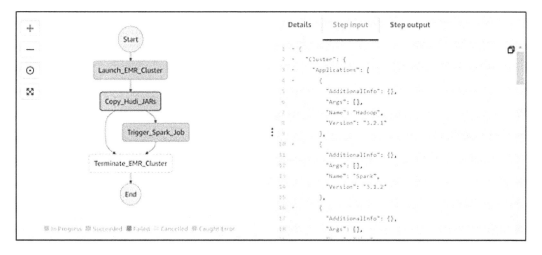

Figure 12.1 – AWS Step Functions visual workflow for EMR jobs

The workflow visualization uses color-coding to represent the status of the tasks. Green indicates that the task has been completed successfully, while blue indicates that the task is currently being executed, and red means that the task failed with an error. While being executed, you can click each task to see its status, the input parameters passed to it, and the output or error it generated following its execution.

After learning how AWS Step Functions can be integrated to trigger EMR jobs, next, let's get an overview of Apache Airflow and MWAA and see how they help to trigger EMR jobs.

Overview of Apache Airflow and MWAA

Apache Airflow is an open source workflow management framework that allows you to build workflows using the **Python** programming language. It has the following fundamental differences compared to AWS Step Functions:

- Being an Apache open source project, Airflow not only supports AWS services, but also supports integration with other public cloud providers and open source projects such as Apache Sqoop, Apache Spark, and many more.

- AWS Step Functions provides a low-code, JSON-based definition, whereas Airflow is more popular with programmers as you need to design a workflow by writing Python scripts.

- AWS Step Functions provides a serverless offering, whereas Airflow needs infrastructure provisioned to act as a cluster on top of which you can run multiple jobs.

From a use case perspective, Airflow is a great fit when *your workflow involves AWS and non-AWS services*. For example, not all your applications are in AWS; a few are on-premises and a few are in another cloud. You would like to build a workflow that invokes an on-premises job, and then triggers an Amazon EMR job and some other cloud job. In this case, AWS Step Functions is not useful as you need to invoke non-AWS jobs and Airflow is a great fit for that. It is very popular in the open source world for designing complex workflows.

To take away the infrastructure provisioning and management overhead, AWS started offering **Amazon Managed Workflows for Apache Airflow (MWAA)**, which is highly available, secure, and is also scalable to serve resource requirements. It is the same as open source Airflow, with support for the same open source integrations, but is a managed offering where you can create a cluster using just a few clicks. It also integrates with AWS CloudWatch and CloudTrail for monitoring and auditing capabilities.

Now, let's see how you can integrate Airflow to invoke EMR jobs using Python.

Integrating Airflow to trigger EMR jobs

Airflow provides the following API functions to interact with the Amazon EMR cluster:

- `EmrCreateJobFlowOperator`: This method enables you to create an EMR cluster.

- `EmrJobFlowSensor`: This helps to check the status of the EMR cluster.

- `EmrAddStepsOperator`: With this, you can add a step to the EMR cluster.

- `EmrStepSensor`: This helps to check the status of an existing step in your EMR cluster.

- `EmrModifyClusterOperator`: This is used to modify an existing cluster.

- `EmrTerminateJobFlowOperator`: This enables you to terminate an existing cluster.

As explained, you can design a workflow in Airflows using the Python programming language, where you can define each action and then define the sequence of execution. The following is sample Python code that executes the `EmrCreateJobFlowOperator` method of Airflow that triggers an EMR create cluster action:

```
cluster_create_action = EmrCreateJobFlowOperator(
    task_id='create_job_flow_task',
    job_flow_overrides=<JOB_FLOW_OVERRIDES>
)
```

You can pass a cluster configuration through the `<JOB_FLOW_OVERRIDES>` parameter that follows a JSON structure.

After a cluster is created, if you need to add a step to the existing EMR cluster, then you can use the `EmrAddStepsOperator` method, as shown in the following sample code, which takes EMR step details through a JSON configuration that is a `<SPARK_STEPS>` variable:

```
add_step_action = EmrAddStepsOperator(
    task_id='add_steps',
    job_flow_id="{{ task_instance.xcom_pull(task_ids='create_
job_flow', key='return_value') }}",
    aws_conn_id='aws_default',
    steps=<SPARK_STEPS>,
)
```

After you have defined all your steps, you can define the sequence of execution by writing something like the following:

```
cluster_create_action >> add_step_action
```

These are a few sample code blocks to explain how you can write Python scripts to define a workflow in Airflow. You can refer to the Airflow documentation to learn more about its methods and way of integration.

Summary

Over the course of this chapter, we have provided an overview of AWS Step Functions, Apache Airflow, and MWAA. In addition, we have shared example code blocks to explain how you can define Step Functions' state machine or write Python code to design a workflow for Airflow.

That concludes this chapter! Hopefully, this helped you get an idea of how you integrate these services to design workflows and will provide a starting point to design more complex data or machine learning pipelines. In the next chapter, we will explain how you can migrate your on-premises Hadoop workloads to Amazon EMR.

Test your knowledge

Before moving on to the next chapter, test your knowledge with the following questions:

1. Assume you are designing a data pipeline that needs to process two input files as two parallel steps and then invoke a common ETL process to aggregate the output of these parallel steps. You have decided to leverage AWS Step Functions to orchestrate the pipeline. Which Task types will you be integrating and how?

2. Assume you have a few Hadoop workloads running on-premises and a few Spark ETL jobs running in Amazon EMR. To simplify orchestration and monitoring, you are looking for an orchestration tool. While comparing different options, you found that AWS Step Functions and MWAA are the two best options. Which of them is better suited to your workload and why?

Further reading

The following are a few resources you can refer to for further reading:

- Learn more about Airflow EMR actions: `https://airflow.apache.org/ docs/apache-airflow-providers-amazon/stable/operators/emr. html`

- Learn more on the integration of AWS Step Functions with Amazon EMR: `https://docs.aws.amazon.com/step-functions/latest/dg/ connect-emr.html`

- Learn how you can leverage AWS Step Functions to orchestrate EMR on an EKS job: `https://aws.amazon.com/blogs/big-data/orchestrate- an-amazon-emr-on-amazon-eks-spark-job-with-aws-step- functions/`

13
Migrating On-Premises Hadoop Workloads to Amazon EMR

Throughout the previous chapters, we have explained what Amazon EMR is, what its features are, how it integrates with AWS services, and how you can integrate a few of the batch or streaming ETL pipelines using EMR. If you are about to start your big data analytics journey, then you can get started with Amazon EMR and other AWS analytics services right away, but there are lot of customers who are already using Hadoop and Spark in their on-premises environments and are in the planning stage to migrate to the AWS cloud.

If you have Hive, Spark, or Hadoop workloads running in an on-premise Hadoop cluster, then there are several factors you need to consider before migrating to AWS, such as support for the Hadoop services you are using, their versions, how security will work in AWS, and what your migration strategy should be.

In this chapter, we will walk through possible migration approaches, options for migrating your cluster data, catalog metadata, ETL jobs, and related workflow services. You will also learn how can you integrate quality control to validate your migration. The following are the high-level topics we will cover in this chapter:

- Migration approaches

- Migrating data and metadata

- Migrating ETL jobs and Oozie workflows

- Testing and validation

- Best practices for migration

Getting an overview of these topics will give you perspectives on the different aspects you should be thinking about while planning for cloud migrations and what changes need to happen to your existing applications to meet the specifications of cloud-native architecture.

Understanding migration approaches

Migrating from an on-premise environment to the AWS cloud provides several benefits including decoupling your compute and storage to provide independent scaling, better security with the AWS infrastructure, the flexibility to design pipelines by integrating other AWS analytics services, and saving resources that would be spent managing infrastructure and instead focusing on application development.

When you plan to migrate your on-premise Hadoop cluster to EMR, you need to analyze how your cluster will work in AWS, compare this with your on-premise environment, and then plan for the migration accordingly. The following are a few of the things you need to analyze:

- Which Hadoop ecosystem services are you using, and are they all supported in AWS?

- If the Hadoop services that are supported are available, then which EMR release version is the closest to your on-premise Hadoop version?

- Does your on-premise cluster use HDFS as a persistent store? When you move to EMR, do you want to continue the same way, or integrate Amazon S3 as the persistent store, which is the recommended approach?

- How much data do you have and how do you plan to migrate it?

- Do you need a single persistent EMR cluster, or you can go with multiple transient workloads?

- How are you managing authentication and authorization on your current on-premise cluster and how do you plan to do that in AWS?

- What are other security aspects of your cluster do you need to consider, such as encryption or networking, and how do you plan to do them in EMR?

- Do you have any centralized metadata catalog, and if so, how do you plan to use it with EMR?

- What kind of ETL jobs do you have, and do they need to be customized to work with S3 in EMR?

- Do you have high-priority jobs that you would like to migrate at the end to avoid any risk affecting your production pipelines?

The answer to all these questions will give you a direction to plan the migration. You will likely be considering the following three approaches:

- **Lift and shift**

- **Re-architecting**

- **Hybrid architecture**

Next, let's understand what each of these approaches means so you can decide on the best approach for your workload.

Lift and shift

Lift and shift means moving the workloads as-is from on-premise environment to AWS without substantial changes to the architecture or ETL scripts, assuming that most of the Hadoop services are available in EMR.

This approach makes more sense in the following situations:

- Your project is time-sensitive and you would like to complete the migration as fast as possible.

- You want to avoid the risk of changes to the architecture or scripts, as it may cause unknown issues while migrating.

- Most of your applications are tied to tight customer **Service-Level Agreements (SLAs)** compared to nightly-batch workloads, so you don't have room to make changes quickly.

Though this approach makes sense in preceding scenarios, there are few prerequisites involved in planning for it, such as the following:

- When you lift and shift your workloads, you need to confirm whether your on-premises big data workloads are using Amazon S3 as persistent storage or a local data node's HDFS. If you are already using S3 as persistent data store, then you can completely avoid data migration and only focus on migrating applications or ETL jobs.

  However, if you are using cluster-node HDFS as the persistent store in your on-premise environment and you are going to embark on a simple lift and shift, then after creating the EMR cluster, you will have to migrate the on-premises cluster HDFS data to the EMR cluster's HDFS so that your ETL scripts continue to work as-is.

- The next thing you should check is which EMR release is closest to your on-premise cluster and whether it has all the required Hadoop ecosystem services at the exact same versions. If not, what minor changes need to be done?

- Also, check for any of the Hadoop services such as Hive, Spark, Tez, and others. What are the default values of the configuration parameters and do you need to override any of them for your workloads?

- EMR clusters use the capacity scheduler by default, so validate whether you can proceed with this or if you'll need to customize the EMR cluster configurations to use Fair Scheduler.

- Check whether you need to integrate additional AWS VPCs, security groups, or IAM roles in order for your authentication, authorization, and connectivity to work as expected.

- Check your on-premise cluster capacity and verify whether you need to have the same or similar capacity using EMR core and task nodes.

Even if you plan to migrate your workloads as-is to EMR, as a next step you should plan to integrate Amazon S3 as your cluster's persistent storage, so that you have the flexibility to integrate multiple transient clusters and also can integrate other AWS analytics services as needed.

Now that we understand the lift-and-shift approach, let's understand how re-architecting might help.

Re-architecting

The primary objective of moving your big data workloads to the cloud is to take advantage of cloud features such as scaling, cost optimization, security, and operational efficiency, among others. If you follow the lift-and-shift approach, then you may not get all the benefits of cloud computing and should look at re-architecting your Hadoop or Spark workloads.

Re-architecting involves a lot of planning and implementation steps, validation of output, and maybe integration with other AWS services. It requires time and resources to re-architect but this is the best way to take advantage of the features of the AWS cloud and Amazon EMR.

The following are some of the aspects you should consider when planning to re-architect your on-premises Hadoop workloads:

- If you are using the Hadoop cluster's HDFS as your persistent storage layer, then plan to migrate your data to Amazon S3 and make application-level changes to work with Amazon S3.

- Check whether your on-premise cluster is being used consistently, meaning whether its CPU and memory resources are used optimally. If not, and all your workloads are batch workloads, then can you consider integrating multiple transient EMR clusters that are job-specific, instead of keeping one long-running cluster for all workloads?

- If your on-premises cluster resources are not used consistently, but you still need a long-running persistent cluster, then check whether starting with a small EMR cluster with managed scaling or scaling with custom policies will help.

- If you are using Hive Metastore as your metadata catalog and plan to build a decoupled architecture with multiple transient EMR clusters, then integrating AWS Glue Data Catalog will add a lot of value by providing a managed and scalable catalog service.

- If you are using Apache Ranger for authorization on your Hive Metastore and plan to re-architect with multiple transient EMR clusters, then integrating AWS Lake Formation for fine-grained access control will provide the flexibility to build a centralized permission management system.

- To orchestrate your transient EMR cluster jobs, you can integrate Amazon EventBridge to trigger jobs based on events or schedules and you can integrate AWS Step Functions to build a workflow.

Apart from the preceding re-architecting best practices, you might have other feature requirements that are missing in EMR and might be planning to build custom workaround solutions for them. Before putting effort in that direction, it's better to check with your AWS Account Manager, who can help by providing information on the EMR feature roadmap and timeline, so that you can take the best decision for your requirements.

Now let's look at how, if you cannot completely re-architect, taking a hybrid approach might also be an option.

Hybrid architecture

You can take a hybrid approach to migration by employing a mix of *lift-and-shift* and *re-architecting*. For existing mission-critical workloads, you can take the lift-and-shift approach, whereas for less critical and new applications you can use the cloud-native re-architecting approach. Then, you can start to slowly migrate mission-critical existing workloads to the AWS cloud-native architecture.

This way, you can avoid the risk of re-architecting mission-critical workloads and will also be able to experiment and gain experience by re-architecting other workloads first.

Now let's see what approaches you can follow for data and metadata catalog migration.

Migrating data and metadata catalogs

As we learned earlier, using Amazon S3 as the persistent data store is the recommended approach when migrating your workloads to AWS or Amazon EMR. If your on-premise environment does not use Amazon S3 as the persistent data store or your existing cluster has Hive Metastore tables, then you need to plan for migrating both data and metadata.

Let's understand what options we have when planning to migrate on-premises cluster data and/or metadata catalogs.

Migrating data

To migrate your on-premises datasets to Amazon S3 or other storage solutions in AWS, you can consider the following tools and services AWS offers:

- Offline data movement using AWS Snowball and Snowmobile, which helps to migrate petabyte- and exabyte-scale datasets.
- For faster online data movement, integrate AWS Direct Connect, which provides dedicated internet bandwidth for data transfers.

- Use Hadoop's `distcp` command to do a distributed copy from on-premises HDFS to S3 using MapReduce.

- Leverage Amazon Kinesis Data Stream or Kinesis Data Firehose to move real-time streaming sources to AWS. You can also leverage AWS DMS to move data from OLTP databases to AWS services.

- Using AWS DataSync or AWS Storage Gateway to migrate files from on-premise environment to AWS.

Now, let's dive deep into each of these approaches.

AWS Snowball and Snowmobile

Both AWS **Snowball** and **Snowmobile** are part of **AWS Transfer Family** and are used to transfer high volumes of data. They also include edge computing capabilities, meaning they offer the CPU capacity to do processing on the data.

If your data volume is so high that it would take months to transfer over the internet, then AWS Snowball or Snowmobile are good options. Let's have an overview of both these services.

AWS Snowball

AWS Snowball is an AWS service using which you can transfer petabyte-scale data from your on-premise environment to AWS. It can transfer multiple terabytes of data and you can cluster multiple devices together to transfer petabyte-scale data in parallel. It can also act as a standalone device for edge computing and is very helpful in places where you have reduced or no connectivity. It is secure as it provides 256-bit encryption and is simple to use.

Snowball Edge provides two options to choose from. One is **Snowball Edge Storage Optimized** and the other is **Snowball Edge Compute Optimized**. Both of them have CPU capacity and block storage, whereas the Compute Optimized version has an optional GPU component for use cases such as machine learning or motion video analysis. You can use these edge computing capabilities in contexts where you don't have any connectivity but you still need compute capacity before the device gets shipped back to AWS.

Snowball has support for AWS Lambda functions and specific EC2 instance types, using which you can develop and test applications and then deploy to devices in remote locations to collect data, do a bit of pre-processing on it, and then finally ship it back to AWS.

Next let's understand, where AWS Snowmobile helps.

AWS Snowmobile

We have learned AWS Snowball can help transfer petabyte-scale data, but what if you have a lot more than that, maybe even exabyte-scale? AWS Snowmobile is a service that can help you transfer exabytes of data from your on-premise environment to AWS. It comes as an actual 45-foot-long container truck that you can connect to your local environment, transfer the data, and then ship the container back to AWS. Each Snowmobile container can transfer 100 petabytes of data.

It is optimized for fast data transfer speeds. As an example, if you have 100 petabytes of data, then with a 1 Gbps data transfer speed, it might take you 20+ years to transfer all of the data. However, with AWS Snowmobile, you might be able to complete the transfer in a few weeks and then ship it back to AWS. Like Snowball, it is also highly secure with 256-bit encryption and the option to integrate AWS KMS to manage the encryption key.

Comparing AWS Snowball with Snowmobile, a single AWS Snowmobile is equivalent as 1,250 Snowball devices. So as a general practice, if you have less than 10 petabytes of data then you can consider Snowball devices, and if you have more than that, consider Snowmobile.

Transferring data with AWS Direct Connect

In the previous section, we discussed large-scale data migrations that would take years to complete over regular internet bandwidth. But if you have terabyte-scale data and you would like to transfer data on an ongoing basis, then using your regular internet connection for data transfers may not be a recommended approach. In such scenarios, to get consistent data transfer speeds using a dedicated network, you can take advantage of AWS Direct Connect.

With AWS Direct Connect, you can establish either one or multiple 1-Gbps or 10-Gbps dedicated connections, using which you can transfer data from your on-premise environment to AWS on a continuous basis. It uses VLANs to access AWS **Virtual Private Cloud** (**VPC**) resources using a private IP address.

Using the S3DistCp utility for data transfer

When you need to copy data from your on-premises Hadoop clusters HDFS directly, then consider the **Apache DistCp** tool or command-line utility, which executes a MapReduce job under the hood to do a distributed copy.

Amazon S3DistCp is an extension of Apache DistCp, which is optimized for transferring data to Amazon S3. You can also use it to load data from Amazon S3 to the EMR cluster's HDFS.

You can use **S3DistCp** as an EMR step too, which will execute a MapReduce job on the cluster for a distributed copy.

The following is a sample Hadoop `distcp` command that copies data from HDFS to Amazon S3 using the `s3a` protocol:

```
hadoop distcp hdfs://<source-hdfs-folder> s3a://<target-s3-path>
```

The following is a sample S3DistCp command that is optimized for Amazon S3:

```
s3-dist-cp --src <source-hdfs-path> --dest s3://<path>
--srcPattern .*\.parquet
```

The optional `srcPattern` parameter allows you to specify a specific file pattern that should be copied alone, instead of all files. In the preceding example, we are going to copy all `.parquet` files from `<source-hdfs-path>` to `s3://<path>`.

> **Important Note**
>
> If S3DistCp failed because it couldn't copy a few specific files, then the EMR step returns a non-zero status and also does not clean up the files already copied. Also, note that S3 bucket names containing underscore characters are not supported, and that S3DistCp cannot concatenate Parquet files.

Migrating real-time streaming sources and on-premises databases

When you plan to migrate your on-premise Hadoop cluster to AWS, your cluster might have real-time streaming workloads or might need other data sources available in a few databases. In those cases, consider integrating Amazon **Kinesis Data Stream** or AWS **Data Migration Service** (**DMS**).

If you have real-time data sources, then as we have learned in previous chapters, you can integrate Amazon Kinesis Data Stream to which your real-time applications can publish data and you can integrate an EMR + Spark Streaming job to read from Kinesis Data Stream and write to the target storage layer.

Similarly, if you have on-premises databases and would like to move data to Amazon RDS Amazon S3, or other supported targets as a one-time extract or **Change Data Capture** (**CDC**), then consider integrating AWS DMS, which is used to move data from a source system to the target system.

Using AWS DataSync to sync data online

AWS DataSync is a secure, managed online data transfer service used to automate and accelerate the movement of data between your on-premise environment and the AWS cloud. It can sync data between different systems including Amazon S3, HDFS, **Network File System** (**NFS**), **Elastic File System** (**EFS**), Amazon **FSx**, and Amazon **FSx for Luster**. It helps to sync your data through your regular internet or AWS Direct Connect dedicated network.

DataSync has features that allow it to scale with data volume, monitor and validate the data transfer, and also notify you in case of failures. All of this makes it a great choice for some of your data stores while you plan for migration.

Using AWS Storage Gateway

AWS Storage Gateway is another data migration service, using which you can sync data between your on-premises filesystems and Amazon S3. You can use this to extend your on-premises storage to the unlimited storage offered by Amazon S3, or you can use Amazon S3 to take a backup of your data.

With this your on-premises environments can access Amazon S3 through an NFS mount point connection, which means anything written to this mount point is automatically synced back to Amazon S3 in its original form. Its low-latency access to Amazon S3 provides interactive file sharing.

This is great when you are planning to migrate in a phased approach, where you plan to sync the data to Amazon S3 first while continuing to use your on-premises systems such as databases and data warehouses.

Migrating metadata catalogs

If you are using Hive Metastore in your on-premise cluster, then in addition to the HDFS data, you also need to consider migrating the Hive Metastore data to AWS. The reason for this is that your existing Hive or Spark jobs can then continue operating on top of the catalog tables defined in your on-premise environment.

When you integrate Apache Hive, then there are few different deployment options to integrate the Hive Metastore such as embedded Metastore, local Metastore, and remote Metastore.

In Amazon EMR, it is strongly recommended to either use AWS Glue Data Catalog as your Metastore or use a remote database such as Amazon RDS as your Hive Metastore. That means you need to plan to migrate your on-premises Hive Metastore data to any of the databases either once or on a continuous basis.

Let's have an overview of how you can migrate the data.

One-time Hive Metastore migration to AWS Glue Data Catalog

When you have an on-premise Hive Metastore and you plan to migrate it to AWS Glue Data Catalog, then you can integrate an AWS Glue ETL job, which will connect to the source Metastore database using a JDBC connection to extract the data and then write into Glue Data Catalog.

This is a one-time migration, so a small application or utility can facilitate a smooth migration. You can look at a few of the sample scripts already available in the AWS Samples GitHub repository. Please check `https://github.com/aws-samples/aws-glue-samples/tree/master/utilities/Hive_metastore_migration`.

One-time Hive Metastore migration to Amazon RDS

Instead of Glue Data Catalog, if you are migrating on-premise Hive Metastore to Amazon RDS, which is a relational database, then you can employ any of the following options to migrate the data:

- You can follow the export and import approach, which means you execute an export job on your on-premises database and import the output in your target Amazon RDS database.

- You can also use database replication features, where any time data is written to on-premises metastore, it will be replicated to the target database.

- You can integrate tools such as AWS DMS or its equivalents to do a one-time full load to the target database.

After migration, you can start using EMR by pointing your Hive Metastore to the new Amazon RDS instance.

Ongoing replication to Amazon RDS

We have learned how you can do a one-time migration of your on-premise Hive Metastore catalog to AWS Glue Data Catalog or Amazon RDS. But you might have specific requirements, because of which you plan to run both your on-premises and Amazon EMR environments for some time and then terminate the on-premise environment. In such scenarios, you will have to keep both the on-premises and Amazon RDS catalogs in sync.

To sync data between both databases, we need to look for the source database changelog, capture the change event, and stream it back to the target database. To make this sync happen, we can think of integrating AWS DMS that supports both one-time data movement and continuous changelog streaming. For our use case here, we can start a DMS task of the type `Migrate existing data and replicate ongoing changes` and make it connect to both the source and target databases to sync them continuously.

The following architecture diagram shows how the flow should look:

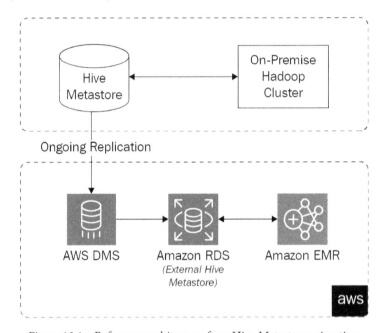

Figure 13.1 – Reference architecture for a Hive Metastore migration

As shown in the diagram, AWS DMS is the centerpiece that migrates the data from the on-premises Hive Metastore to Amazon RDS.

In this section, we have learned about both data and metadata migration. As a next step, we should review ETL jobs being executed in your on-premise cluster and plan for their migration.

Migrating ETL jobs and Oozie workflows

If you are doing lift and shift and your ETL scripts are configured to read from and write to HDFS, then your existing ETL scripts such as Hive, MapReduce, and Spark will work just fine in EMR without substantial changes. But if, while migrating to AWS, you re-architected to use Amazon S3 as your persistent layer instead of HDFS, then you will have to change your scripts to interact with Amazon S3 (s3://) using EMRFS.

> **Important Note**
> Prior to the release of Amazon EMR 5.22.0, EMR supported the `s3a://`
> and `s3n://` prefixes to interact with EMRFS. These prefixes haven't been
> deprecated and still work, but it is now recommended to use the new `s3://`,
> which provides a higher level of security and easier integration with Amazon S3.

Apart from your Hive and Spark scripts, if you are using Apache Oozie for workflow
orchestration of your ETL jobs, then you need to plan for its migration too. Let's
understand what options you have for this.

Migrating Oozie workflows

Apache Oozie is a *workflow scheduler* that is very popular in the Hadoop ecosystem to
manage and orchestrate Hadoop jobs such as Hive, Pig, Sqoop, Spark, DistCp, Linux
shell actions, and many more. It is a scalable, reliable system that is popular in the
Hadoop world.

Oozie has two components: one is *workflow jobs*, using which you can design workflow
steps as **Directed Acyclic Graphs** (**DAGs**). The other is **Oozie Coordinator**, which is used
to schedule your workflow jobs to run with an event or through timed schedule.

Oozie allows you to define workflows using XML definitions and is available in Amazon
EMR starting with the 5.0.0 release.

Similar to Hive, Oozie also has a Metastore database for which you will need to plan the
migration. When considering migrating Oozie workflows to EMR, we need to migrate all
workflow definition files and the Metastore database.

Migrating Oozie Metastore databases

By default, Oozie uses **Apache Derby** as its Metastore database and also provides a
command-line option to export its database. Similar to Hive Metastore in EMR, it
is recommended to keep the Oozie Metastore outside of the EMR cluster for better
reliability, so consider using Amazon RDS as the Oozie Metastore.

To migrate an Oozie database to the Amazon RDS MySQL engine, refer to the following
steps that guide you through the export and import process:

1. Log in to the Oozie server node of your on-premise cluster, navigate to the path
    where `oozie-setup.sh` exists, and execute the following Oozie command to
    export the Metastore database:

    ```
    ./oozie-setup.sh export /<path>/<oozie-exported-db>.zip
    ```

    In EMR, `oozie-setup.sh` can be found in `/usr/lib/oozie/bin/`.

2. Next, upload the exported database ZIP file to Amazon S3, from which you can import:

```
aws s3 cp <oozie-exported-db>.zip s3://<bucket-name-
path>/<oozie-exported-db>.zip
```

3. Then SSH into the EMR master node using PuTTY or equivalent software by following the steps explained in the *PySpark Notebook* section of *Chapter 11, Implementing UPSERT on S3 Data Lake with Apache Spark and Apache Hudi*, and then execute the following step to download the exported Oozie database file from Amazon S3 to the local filesystem:

```
aws s3 cp s3://<bucket-name-path>/<oozie-exported-db>.zip
<oozie-exported-db>.zip
```

4. Next, you need to create the Oozie database in Amazon RDS and grant the required permissions. Log in to your database as root and execute the following commands in the MySQL prompt:

```
mysql> create database oozie default character set utf8;

mysql> grant all privileges on oozie.* to
'oozie'@'localhost' identified by 'oozie';

mysql> grant all privileges on oozie.* to 'oozie'@'%'
identified by 'oozie';
```

5. After you have created the database with all the required permissions, as a next step we can import the database file using the same `oozie-setup.sh` utility as shown in the following command:

```
./oozie-setup.sh import <oozie-exported-db>.zip
```

6. After the database is ready with all the imported metadata, we need to make some changes to the Oozie configuration in EMR, so that it points to the new Amazon RDS database. Please make the following changes in the `oozie-site.xml` configuration file:

```
<property>
    <name>oozie.service.JPAService.jdbc.driver</name>
    <value>com.mysql.jdbc.Driver</value>
</property>
<property>
```

```
        <name>oozie.service.JPAService.jdbc.url</name>
        <value>jdbc:mysql://<amazon-rds-host>:3306/oozie</
value>
    </property>
    <property>
        <name>oozie.service.JPAService.jdbc.username</name>
        <value><mysql-db-username></value>
    </property>
    <property>
        <name>oozie.service.JPAService.jdbc.password</name>
        <value><mysql-db-password></value>
    </property>
```

Please make sure you replace the `<amazon-rds-host>`, `<mysql-db-username>`, and `<mysql-db-password>` variables in the preceding code with your own database connection information.

7. As a final step, to reflect the changes in Oozie, we need to restart the Oozie service using the following command:

```
sudo restart oozie
```

These steps help in migrating the Oozie Metastore database to the remote EMR RDS database. As the next step, you need to move all the workflow definition files to EMR.

Migrating Oozie workflow definitions

For every workflow definition, you will have an XML definition file and the following dependent files:

- `job.properties`
- `workflow.xml`
- `coordinator.xml`
- Any other dependent file

You can take a backup of these files by archiving them as a single ZIP file, uploading it to Amazon S3, and then copying it back to EMR using the `aws s3 cp` command for integration. Make sure, to modify the workflow configuration file to reflect the connection with EMR.

Then you can submit your jobs to EMR as you used to do in your on-premise environment, and can use the Hue interface to view your Oozie workflows.

The next step of your migration is testing and validation to make sure your cluster setup works as expected and that the data was migrated accurately.

Testing and validation

In the previous section, we learned how we can migrate data, metadata, ETL jobs, and workflows, but after the migration is complete, it's very much essential to validate the migration with a proper testing strategy.

Your options for data validation will vary based on the methodology you used for your migration. We previously explained the different phases of a migration where you migrate the data and metadata separately. So, let's now understand how you can validate the data quality for each of those phases.

Validating metadata quality

We discussed about migrating Hive and Oozie Metastore, and you can apply same knowledge to migrate Hue Metastore too. All of them integrate a relational database as their Metastore, which means we have the option of executing standard SQL statements to count records or validate data.

Let's look at a few of the options we can consider to validate our metadata migration:

- **Relational data migration with AWS DMS**: If you used AWS DMS for any relational database migration, then you can look at the table statistics provided by AWS DMS that shows how many rows were added, updated, and deleted per table. In addition, DMS also provides options to create control tables in the target system, which compares the source and target table records to validate whether the data migration was successful. This is helpful for when we plan to migrate on-premises Hive Metastore tables to Amazon RDS.

- **Manually running queries**: As the metadata volume is not as huge as data (usually in the GB to TB range), we can also include manual testing as one of the approaches to validate the quality of data. This involves running a few standard SQL queries to confirm the table record count or field population.

- **Directly integrating services with the cluster**: This is another way you can validate a successful migration of Metastore, where you use the respective services such as Hive, Spark, Hue, or Oozie to validate that they work as expected. For example, are you able to create a new Hive database or table? Are you able to run queries on a Hive table using Spark code?

These are few of the options you can look at, and you also can explore other third-party products (both free and paid) that can help validate the metadata as well. Next, let's understand how you can validate the data itself.

Validating data quality

Now that we understand our options for validating our metadata migration, let's next see what options we have to validate the actual data that we transferred. This is not an exhaustive list, rather a few of the options you can consider.

Amazon S3 Inventory Report

Amazon S3 Inventory Report is a great feature of Amazon S3 that you can enable on your S3 bucket or a specific folder to generate a report in CSV, ORC, or Parquet formats. This report includes stats about objects, storage classes, encryption, and more. You can select which attributes you need as part of your report and schedule it to be generated every day or week.

Enabling S3 Inventory Report is very easy and offers a great way to validate your data. If you have transferred data from HDFS to Amazon S3, then the bucket or folder size could help in identifying whether all the data was copied successfully. You can also use the object count if the files were transferred directly, instead of via MapReduce or Spark jobs, as they may create different set of `PART` files, named `part-0`, `part-1... part-n` in the S3 target path.

Sqoop or DistCp job output

If you used a **Sqoop** job to transfer the data, or used the DistCp command, which employs a MapReduce job under the hood, then you can rely on the command output to validate its success.

Sqoop provides a `-validate` option in its CLI that instructs Sqoop to compare the source record count with that of the target to validate the success of the transfer. The following is a sample Sqoop `export` command with the `-validate` option:

```
sqoop export -connect jdbc:mysql://<host>/<db-name> -table
<table-name> -export-dir /<path> -validate
```

Similarly, for the DistCp command, you can specify the `-log <log-dir>` parameter to analyze the logs to validate the success of command execution or data transfer.

Deequ framework

You might be familiar with the unit test cases that we write for application scripts. An example of this is JUnit, which is integrated into Java applications to do unit testing on your code. It would be great if we had similar tool or utility to test our data quality too, which could validate whether the data quality is good. This is where the Deequ framework comes in and adds lots of value. Deequ is used internally in Amazon and also by many customers to validate the quality of their terabyte-scale datasets with the Spark processing engine.

Deequ is a GitHub-based utility, built on top of Apache Spark and used to process large volumes of data for quality checking. It has the following three high-level components:

- **Metrics Computation**: Deequ performs data quality checks with a set of statistical metrics including the maximum value, minimum value, correlation, and completeness of a column.

- **Constraint Verification**: Deequ provides the flexibility to define constraints on your data and it uses same constraints to derive metrics and generate a report that includes constraint verification results.

- **Constraint Suggestion**: Deequ has intelligence built in to analyze your data and suggest constraints that you can implement.

The following is a high-level architecture diagram that represents the Deequ components:

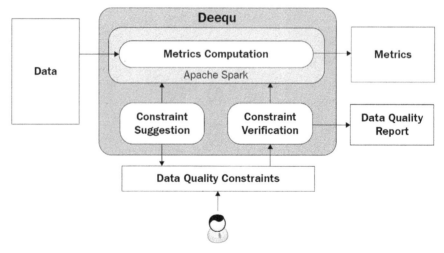

Figure 13.2 – Reference architecture for Deequ

Deequ provides several metrics while validating the data quality. Some examples of the popular metrics are `Completeness`, `Compliance`, `Correlation`, `CountDistinct`, `Distinctness`, `Maximum`, `Mean`, `Minimum`, `Sum`, `UniqueValueRatio`, and `Uniqueness`.

There is a Python equivalent of Deequ called **PyDeequ**, using which you can integrate Deequ functionality in PySpark code and AWS Glue jobs.

AWS Glue DataBrew profiling

AWS Glue DataBrew is a service built on top of the Spark framework for distributed processing of large volumes of data. It provides a rich visual interface and offers 250+ built-in transformations that you can use to build your ETL flows, all without writing any code.

AWS Glue DataBrew also provides a feature to profile your data, which you can use to check your data quality. This is very useful, as without writing any code, you can profile your data in the visual interface by specifying a list of columns you want to profile. The profile output includes several metrics such as `correlation`, `duplicate row count`, `distinct`, `max`, `mean`, `median`, `mode`, `percentile`, `range`, `sum`, and `variance`.

With additional ETL transformations, you can automate the conversion of your data to standard format, fixing missing or invalid values and also filtering anomalies. Similar to most AWS services, AWS Glue DataBrew also offers a pay-as-you-go model, which means you pay for node usage for the duration of the job execution.

Manual sanity check

In all previous options we have discussed, what are some of the tools you can use to check data quality. Even if you have implemented automation throughout the process, it's still a good practice to do a bit of manual checking, either by opening a few files or by running a sample ETL job that you know ran without problems in your on-premise environment.

We have now covered how you can validate your migration and check the quality of the migrated data and metadata with a variety of tools. In the next section, we will cover some of the best practices to follow while migrating your workloads to AWS.

Best practices for migration

The following are some of the best practices you should follow when onboarding your solutions into a cloud-native architecture:

- **Split batch and interactive or streaming workloads**: Look for opportunities to build transient EMR workloads so that your persistent cluster resources are not idle when you don't have any processes running. Of course, you might have other workloads where a persistent cluster is required, such as for interactive development or real-time streaming workloads, so it's better to identify which workloads need the persistent cluster, and then move the other workloads to transient job-specific EMR clusters.

- **DevOps automation**: For the launching of clusters or other AWS resources, consider integrating AWS CloudFormation to automate the creation of the required infrastructure resources. This increases efficiency when you plan to launch the same set of resources and configurations in multiple environments, such as development, staging, and production. For **Continuous Integration** (**CI**) and **Continuous Deployment** (**CD**) pipelines, you can either continue with your on-premise setup or integrate the AWS CodeCommit, CodeDeploy, or CodePipeline services, which provide CI/CD capability on AWS.

- **Reserved and Spot instances**: Reserved and Spot instances provide cost savings. If you have defined workloads that require specific EC2 instance types for years to come, then you can purchase **Reserved instances**, which provide good discounts for usage commitments.

  On the other hand, when integrating autoscaling for your EMR cluster, using EC2 **Spot** instances of your task nodes provides great discounts. Amazon EC2 Spot Instance is an alternate purchasing option compared to EC2 on-demand type, which allows you to purchase EC2 capacity that is unused and is available for purchase with higher discount. You can acquire EC2 Spot instances through bidding process but they come up with a risk of getting terminated at any point in time as the price goes beyond your bidding price.

- **Regulatory and compliance requirements**: Check whether your organization has specific regulatory or compliance requirements and whether EMR meets all your needs in this regard. You can identify these requirements from `https://docs.aws.amazon.com/emr/latest/ManagementGuide/emr-compliance.html`.

- **Skill gaps**: When migrating from on-premise to the AWS cloud, you need to identify your team's skill levels with AWS and EMR to see how they can help with the migration. If they are new to AWS, you can offer them the necessary training resources to upskill them to handle the migration project.

- **Prototyping**: Before migrating your workloads, building a small prototype using the EMR cluster and integrating other related services is a good idea. This allows you to explore any unknowns and gain the confidence to proceed with the migration.

- **Cost estimation**: It is recommended that you calculate the **Total Cost of Ownership (TCO)** in AWS and look at all aspects of the application before migrating to AWS. Firstly, define the number of transient and persistent EMR clusters you require, the size of the clusters, the types of EC2 instances you will be using, and for how long the clusters will be active. This should give you a ballpark estimation of your EMR cost. On top of that, factor in Amazon S3, AWS Glue Data Catalog, the Metastore databases, and any other optional AWS services you will be integrating into your architecture. The AWS Pricing calculator available at `https://calculator.aws/#/` is a great tool for estimating your cost.

- **Define data retention policies**: When using Amazon S3 as your persistent data store, you have several options to control your data accessibility. You can choose from Amazon S3's different storage classes such as S3 Standard, S3 Intelligent Tiering, S3 Standard-IA, and Glacier depending on your usage pattern. For example, if some of your data is no longer used for processing, then you can create S3 life cycle polices to move old data to Glacier, so that you can retain data longer with reduced storage costs. Glacier is a type of S3 storage class, which is used to archive data with an assumption that you will not be accessing your data frequently and you pay a minimal fee to retrieve data.

These are few of the general best practices that should help you to get started. Moving forward, you can dive deep into each of the workloads and apply the knowledge you have gained from the previous chapters to plan your migration.

Summary

Over the course of this chapter, we have looked at an overview of different migration strategies you can follow while migrating your on-premises Hadoop workloads to AWS, and how you can migrate data, metadata, and ETL jobs. We then covered a few testing and validation strategies you can follow to check the quality of your data, and also discussed some of the best practices you can follow during the migration process.

That concludes this chapter! Hopefully, this helped you get an idea of how you can plan your migration and some of the aspects you should be considering. In the next chapter, we will examine some of the best practices for EMR, along with how you can optimize costs while integrating your ETL flows.

Test your knowledge

Before moving on to the next chapter, test your knowledge with the following questions:

1. Assume you have several on-premises Hadoop workloads, out of which a few are subject to sensitive customer SLAs, and your organization has decided to move all workloads to AWS. Which migration strategy do you think is ideal for your use case?

2. Assume you have around 100 petabytes of data in your on-premise environment and you are planning to migrate the data to Amazon S3. Looking at the volume of data, which data migration strategy or tool do you think is best for your use case?

3. Assume you have completed the migration of your on-premise environment that included several Hadoop workloads and 100s of terabytes of data. Now you are looking for ways to validate the data quality in Amazon S3. Which tool or utility will be helpful to check the quality of the data?

Further reading

The following are a few resources you can refer to for further reading:

*   Amazon EMR Compliance: `https://docs.aws.amazon.com/emr/latest/ManagementGuide/emr-compliance.html`

*   AWS Risk and Compliance whitepaper: `https://docs.aws.amazon.com/whitepapers/latest/aws-risk-and-compliance/welcome.html`

*   A look at the EMR migration program offered by AWS: `https://aws.amazon.com/emr/emr-migration/`

*   A look at AWS Data Lab, which can help to architect your solution in AWS: `https://aws.amazon.com/aws-data-lab/`

*   Data profiling with AWS Glue DataBrew: `https://docs.aws.amazon.com/databrew/latest/dg/jobs.profile.html`

*   The Deequ framework for data quality checks: `https://github.com/awslabs/deequ`

*   Amazon S3 Intelligent Tiering: `https://aws.amazon.com/s3/storage-classes/`

14

Best Practices and Cost-Optimization Techniques

Welcome to the last chapter of the book! During all the previous chapters, you learned about EMR and its advanced configurations and security. You also learned how you can migrate your on-premise workloads to AWS and how you can implement batch, streaming, or interactive workloads in the AWS cloud. In this chapter, we will focus on some of the best practices and cost optimization techniques you can follow to get the best out of Amazon EMR.

When considering best practices for implementing big data workloads in EMR, we should look at different aspects such as EMR cluster configuration, sizing your cluster, scaling it, applying optimization on S3 or HDFS storage, implementing security best practices, and different architecture patterns. Apart from these, optimizing costs is also a best practice and AWS provides several ways to optimize costs and offers various tools to monitor, forecast, and get notified when your spending goes beyond the defined budget threshold.

The following are the high-level topics we will be covering in this chapter:

- Best practices around EMR cluster configurations
- Optimization techniques for data processing and file storage
- Security best practices
- Cost-optimization techniques
- Limitations of Amazon EMR and possible workarounds

As we saw in *Chapter 13*, *Migrating On-Premises Hadoop Workloads to EMR*, migrating to AWS or using AWS resources provides a lot of benefits, but understanding how you can get the best out of those AWS services and how you can optimize your workloads on AWS adds a lot of value as you truly get the power of the AWS cloud. Apart from best practices, it is also important to understand different limitations, so that you can plan for different workarounds.

Best practices around EMR cluster configurations

When you start using EMR clusters for your Hadoop workloads, the primary focus is on writing logic for the ETL pipeline, so that your data gets processed and is available for end user consumption. There are several factors you might have considered to optimize your ETL code and the business logic integrated into it but, apart from optimizing code, there are several other optimizations that you can consider in terms of EMR cluster configurations that can help optimize your usage.

Let's understand some of the best practices that you can follow.

Choosing the correct cluster type (transient versus long-running)

As explained in *Chapter 2*, *Exploring the Architecture and Deployment Options*, there are two ways you can integrate EMR clusters. One is a long-running EMR cluster, which is useful for multi-tenant workloads or real-time streaming workloads. Then we have short-term job-specific transient clusters that get created when an event occurs or on a schedule, and then get terminated after the job is done.

When you have batch workloads, there will be some jobs where you need a fixed-capacity cluster. Then there might be a few other workloads where the cluster capacity you need will be variable and you will have to take advantage of EMR's cluster scaling feature.

So, looking at your workload, selecting the type of EMR cluster is the first best practice you can follow. The following diagram shows all three types of implementations you can choose:

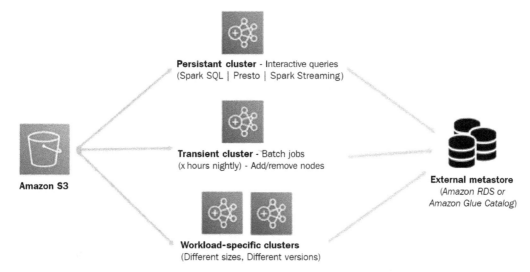

Figure 14.1 – Reference architecture for multiple EMR workloads sharing S3 and a metadata catalog

As you can see, all the EMR clusters are sharing the same external metastore and Amazon S3 persistent store. Let's understand how it helps using Amazon S3 with EMRFS and what value is added with an external metastore.

Choice of storage (HDFS versus Amazon S3)

While implementing multiple transient EMR clusters, Amazon S3 plays a crucial role as all transient EMR clusters share data using S3 and you can also leverage other AWS services such as AWS Glue, Amazon Athena, or Amazon Redshift to access and process the same datasets. With S3, you get higher reliability and also can integrate different security and encryption mechanisms.

While Amazon S3 is widely used for most big data processing use cases, still there are some use cases where HDFS storage is needed for faster querying. For those workloads, you can think of initiating your cluster with a DistCp step that copies data to your EMR cluster for long-running workloads and then syncs it back to Amazon S3 so that your other EMR clusters or AWS Glue-based jobs can read the same datasets.

Externalizing the metastore

As we learned in *Chapter 4, Big Data Applications and Notebooks Available in Amazon EMR*, externalizing a cluster metastore such as a Hive or Hue metastore is one of the best practices as it enables you to split your workloads into multiple transient workloads and also provides higher reliability.

When you externalize your Hive metastore, integrating AWS Glue Data Catalog is recommended as that is a managed scalable service and also enables you to integrate other AWS services such as Amazon Redshift, AWS Glue, Amazon Athena, Amazon QuickSight, or AWS Lake Formation in the same centralized catalog and provides unified data governance.

In this section, we have looked at best practices around cluster type, metadata catalogs, and persistent storage. Next, let's see what best practices we can follow while configuring an EMR cluster.

Best practices around sizing your cluster

When you are configuring an EMR cluster, there are a few best practices that you can follow to get the best out of it. The following are some of the priority ones, but there can be others, depending on the workload.

Choosing the right instance type

When you are configuring an EMR on EC2 cluster, you have flexibility to select the type of instance you will be integrating for the master as well as core or task nodes. Depending on your workloads, you should choose the right EC2 instance type.

Choosing the master node instance type

The master nodes in Hadoop or EMR clusters do not process ETL jobs or tasks, rather they act as driver nodes to coordinate between core, task nodes and also keep track of cluster and node status. For master nodes, its best to choose either EC2 M5, C5, or R5 instances depending on the use case:

- If you have fewer than 50 core and task nodes, then we can consider that a low or medium cluster and a master node of `m5.xlarge` instance type will be able to manage the cluster jobs.

- If you have more than a 50-node cluster, then the next thing you can check is whether you have workloads that have a heavy network I/O. If yes, then the C5 or R5 instance family with enhanced networking is the recommended option. If you don't need enhanced networking, then you can choose higher than the `m5.xlarge` instance type.

These are general guidelines to give you a starting point. Then, depending on your use case, which might have higher read versus write or a higher number of jobs running in parallel, you need to experiment a bit with different instance types before making a decision.

Choosing the core and task node instance type

For core or task nodes, you need to first understand and identify the purpose of your EMR cluster. Then, based on that, you can select any of the following:

- **General Purpose**: In general, batch ETL workloads fall into this category, which can integrate M4 or M5 EC2 instance family nodes.

- **Compute Intensive**: You might have compute-heavy workloads such as machine learning jobs that need more CPU power. For those workloads, you can consider integrating C4 or C5 instances.

- **Memory Intensive**: For memory-intensive workloads such as interactive analysis or fast querying use cases, you can consider integrating R5 or X1 instances that have higher memory capacity.

- **Dense Storage**: If you have workloads that require large HDFS capacity, which means every core node should have more disk capacity, then you can consider D2 or I3 EC2 instance family nodes.

These instance type suggestions are based on the type of EC2 instances available while writing this book. Please refer to the latest supported instance types while configuring your EMR cluster.

Deciding the number of nodes for your cluster

After selecting the correct instance type for your use case, the next question we need to answer is how many instances we should configure for our cluster.

As you might be aware, for Hadoop workloads, every input dataset gets divided into multiple splits and then gets executed in parallel using multiple core or task nodes. If you have configured a bigger instance type, then your job will have higher memory and CPU capacity, which might help you to complete the job faster. On the other hand, if the cluster has smaller instance types, then your mapper or reducer tasks wait for the resources to be available and take a comparatively longer time to complete the job.

All your mappers can run in parallel, but that does not mean you can add unlimited nodes to your cluster so that your mappers complete without having to be queued. But in reduce or aggregate operations, most of the nodes go into the idle state, so you have to experiment and arrive at the right number of nodes. In addition, you can try answering the following questions, which can guide you to arrive at the correct number of instances to start with:

- **How many tasks will you have to execute?**

  Look at the input file size, using which you can guess the file split size, and then look at the operation you are going to do to have a rough guess of the number of total tasks you might need.

- **How soon do you want the job to be completed?**

  This question will help you evaluate how much parallelism you should be aiming at and also how many tasks each of the core or task nodes can handle based on the CPU or memory the job or task might have.

- **How many core or task nodes might you need?**

  HDFS and Hadoop services run only on the core nodes. So, depending on the HDFS size you need, you can select the number of core nodes and configure the rest as task nodes where you can implement auto-scaling. Please try to maintain a 1:5 ratio between core and task nodes, which means do not add more than five task nodes per single core node.

In general, one of the recommendations you can follow is to prefer a smaller cluster with a larger instance type as that can provide better data locality and your Hadoop or Spark processes can avoid more shuffling between nodes.

Determining cluster size for transient versus persistent clusters

Depending on whether you have a transient or persistent cluster, the following are some of the sizing best practices you can follow:

- **Transient cluster use cases**: For transient EMR cluster use cases, which are mostly used for batch ETL workloads, if you do not have strict **Service-Level Agreements (SLAs)**, then think of using **Spot EC2** instances with task nodes, so that you can save costs. If you have strict SLAs to meet, then you can consider using on-demand or reserved instances for predictable costs.

- **Persistent cluster use cases**: For persistent clusters, think of using **Reserved Instances** for master and core nodes. For scaling needs, use **Spot** or **On-Demand instances**, depending on your SLA requirements.

After understanding how you should choose cluster instance type and the number of nodes, next we will explain how you can configure high availability for your cluster's master node.

Configuring high availability

For high availability, you should configure multiple master nodes (three master nodes) for your cluster, so that your cluster does not go down when a single master node goes down.

> **Important Note**
>
> All the master nodes are configured in a single **Availability Zone** and in the event of failover, the master node cannot be replaced if its subnet is overutilized. So, it is recommended to reserve the complete subnet for the EMR cluster and also make sure the master node subnet has enough private IP addresses.

Apart from master nodes, if you need to enable high availability for core nodes, then consider using a core node instance group with at least four core nodes. If you are launching a cluster with a smaller number of core nodes, then you can think of increasing the HDFS data replication by setting it to two or more.

Best practices while configuring EMR notebooks

In *Chapter 4*, *Big Data Applications and Notebooks Available in Amazon EMR*, we explained the usage and benefits of EMR notebooks, which you can use for interactive development and attach them to an EMR cluster for job execution.

The following are some of the best practices you can follow while integrating EMR notebooks:

- It's better if you keep the notebooks outside of the cluster so that you can attach them to or detach them from different clusters as needed.
- You can configure multiple users to attach their notebooks to the same cluster. You can also enable auto-scaling on the clusters to support them with the required resources.
- Configure to save notebooks to Amazon S3 for better reliability.
- Integrate GitHub control for code sharing.
- Enable tag-based access control.

Next, we will explain how Apache Ganglia can help in monitoring your cluster resources.

Using Ganglia for cluster usage monitoring

As explained in *Chapter 4, Big Data Applications and Notebooks Available in Amazon EMR*, Apache Ganglia is an open source project, which is scalable and designed to monitor the usage and performance of distributed clusters or grids.

In an EMR cluster, Ganglia is configured to capture and visualize Hadoop and Spark metrics. It provides a web interface where you can see your cluster performance with different graphs and charts representing CPU and memory utilization, network traffic, and the load of the cluster.

Ganglia provides Hadoop and Spark metrics for each EC2 instance. Each metric of Ganglia is prefixed by category, for example, distributed filesystems have the `dfs.*` prefix, **Java Virtual Machine** (**JVM**) metrics are prefixed as `jvm.*`, MapReduce metrics are prefixed as `mapred.*`.

If you have a persistent EMR cluster, then Ganglia is a great tool to monitor the usage of your cluster nodes and analyze their performance.

Tagging your EMR cluster

Tagging your AWS resources is a general best practice, which you can apply to Amazon EMR too. Every time you create a cluster, it's better to provide as much metadata as possible using tags such as project name, team name, owner, type of workload, and job name.

For example, you can identify EC2 instances that are part of your EMR cluster using the following tags:

- `aws:elasticmapreduce:instance-group-role=CORE`
- `aws:elasticmapreduce:job-flow-id=j-<id>`

You can use these tags for reporting, analytics, and controlling costs too. It is recommended you arrive at a set of tag keys first and use it consistently across clusters.

As a best practice, do not include any sensitive information as part of your tag keys or values as they are used by AWS for reporting.

Optimization techniques for data processing and storage

We have recommended using Amazon S3 as the EMR cluster's persistent storage as it provides better reliability, support for transient clusters, and it is cost-effective. But there are several best practices we can follow while storing the data in Amazon S3 or an HDFS cluster.

Let's understand some of the general best practices that you can follow to get better performance and save costs from a storage and processing perspective.

Best practices for cluster persistent storage

As part of your cluster storage, there are some general best practices that apply to both Amazon S3 and HDFS cluster storage. The following are a few of the most important ones.

Choosing the right file format

You might be receiving files in CSV, JSON, or as TXT files, but after processing through the ETL process, when you write to a data lake based on S3 or HDFS, you should choose the right file format to get the best performance while querying it.

We can divide the file formats into two types – *columnar* or *row*-based formats. Row-based formats are good for write-heavy workloads, whereas columnar formats are great for read-heavy workloads.

Most of the big data workloads are related to analytics use cases, where data analysts or data scientists perform column-level operations such as finding the average, sum, or median of a column. Because of that, for analytical workloads, columnar formats such as **Parquet** are very popular.

Apart from the columnar nature, you should also be considering whether the file format is *splittable*, which means can the Hadoop or Spark framework split the file for parallel processing. In addition, we should check whether the file format supports schema evolution, which means whether incremental files can support additional columns.

The following table shows a comparison of features supported by the text, ORC, Avro, and Parquet file formats:

Features	Parquet	ORC	Avro	Text
Supported in EMR	Yes	Yes	Yes	Yes
Block Compression	Yes	Yes	Yes	No
Schema Evolution	Yes	Yes	Yes	No
Data Storage	Column	Column	Row	Row
Write Performance	Slow	Slow	Medium	Fast
Read Performance	Fast	Fast	Medium	Slow

Figure 14.2 – Table comparing features of different file formats

These four formats are commonly used and by looking at their features, you can choose the right file format for your use case.

Choosing the compression algorithm

On top of the correct file format, you can also apply an additional compression algorithm to get improved performance for file transfers. While choosing a compression algorithm, we also need to make sure the compression algorithm is splittable so that it can be split by Hadoop or Spark frameworks for parallel processing.

Each compression algorithm has a rate of compression that reduces your original file size by x percentage. The higher the compression rate, the more time it takes to decompress. So, it's a space-time trade-off, where you save more storage with a high compression rate and spend more time decompressing it.

The following table shows a comparison of the popular compression algorithms.

Algorithm	Splittable?	Compression Ratio	Compression / Decompression Speed
Gzip	No	High	Medium
Bzip2	Yes	Very High	Slow
LZO	Yes	Low	Fast
Snappy	No	Low	Very Fast

Figure 14.3 – Table comparing compression algorithms

This should help you make a decision on the file format and compression algorithm you should integrate.

Choosing an S3 storage class

If you are using Amazon S3 as your cluster's persistent storage, then you have additional flexibility to choose from different S3 storage classes. S3 standard is the commonly used storage class, assuming you are accessing your data frequently. But if you have data that is not being frequently accessed and you would like to save costs, then you should move your data to any of the following S3 storage classes:

- S3 Standard-IA
- S3 One Zone-IA
- S3 Glacier Instant Retrieval
- S3 Glacier Flexible Retrieval
- S3 Glacier Deep Archive
- S3 Intelligent Tiering

Out of all the preceding options, S3 Intelligent Tiering can be a good option to choose if your access patterns are not fixed and you are not sure when to move to another storage class. S3 Intelligent Tiering looks at your access pattern and automatically moves objects to the most cost-effective storage tier without any effect on performance or without any retrieval costs or operational overhead. It provides high performance for **Frequent**, **Infrequent**, and **Archive Instant Access** tiers.

Data partitioning

This is applicable to both Amazon S3 and HDFS storage, where you need to structure your data into folders and subfolders so that your queries perform better. Assume you are receiving weather data every day and your query patterns on them are mostly date-based filters. In such use cases, if you store your data with the `<year>/<month>/<date>` sub-folder structure and you write SQL queries with `WHERE year=<value> AND month=<value>`, it will scan the respective sub-folder to get the data instead of scanning all the folders.

In S3, the path will look like this: `s3://<bucket-name>/<year>/<month>/<day>/`.

Best practices while processing data using EMR

While processing your data in EMR, you have the option to use Hive, Spark, Tez, or any other Hadoop frameworks for your ETL or streaming workloads. Based on the framework you are using, you can apply its tuning parameter to get the best performance. For example, if you are using Spark, you can play with the Spark executor or the driver's memory and CPU parameters to get the best performance, or you can pass any other tuning parameters open source Spark offers.

Security best practices

Security is one of the important aspects when you move to the AWS cloud. It includes authentication, authorization on cluster resources, protecting data at rest and in transit, and finally, protecting infrastructure from unauthorized access. We have discussed these topics in detail in *Chapter 7, Understanding Security in Amazon EMR.*

The following are a few of the general best practices that you can follow while implementing security:

- Follow the least privilege principle of AWS and provide the minimal required access to your cluster.

- Avoid using the same AWS IAM role for multiple clusters; rather, create use case or cluster-specific roles to reduce the blast radius.

- If you do not have a specific EMR release dependency, then prefer to use the latest EMR release, which has all the security patches integrated.

- It's better to consider all security aspects from the very beginning, as implementing it later is more complex and expensive.

- Continuously review your organization's security guidelines and review your implementation in AWS and Amazon EMR.

- Use EMR security configuration to templatize the setup and apply it to multiple clusters.

- It's better if you launch your clusters within a private subnet so that you can limit access to your cluster.

- Leverage AWS CloudFormation to create cluster resources so that you can use the application to create resources on other lower or higher environments.

Apart from the general security best practices, you can consider a few of the following specific best practices to make your environment more secure.

Configuring edge nodes outside of the cluster to limit connectivity

When you have a persistent EMR cluster and you plan to provide SSH access to it, or you would like to configure port forwarding to access the Spark history server or Ganglia web UI, then it is recommended to create a separate edge node, instead of providing access to the EMR cluster's master node.

Not only for a single cluster but if you have multiple EMR clusters, then you can have a common edge node in a public subnet of your VPC to limit access to your cluster nodes.

The following is a reference architecture that shows how you can configure a common edge node in a public subnet, which can be used as a jump box to connect to EMR clusters available in private subnets.

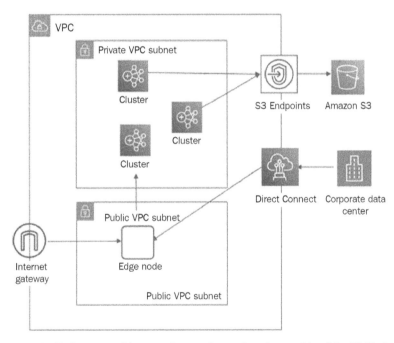

Figure 14.4 – Reference architecture for an edge node to be outside of the EMR cluster

This also provides another recommendation, which is to configure an S3 endpoint to access Amazon S3, which can avoid the request routing through the public internet and gets better performance with access through an Amazon internal network.

You can connect to the edge node from your corporate data center using Direct Connect, or can directly access it through an internet gateway as the edge node is available in the public subnet.

Integrating logging, monitoring, and audit controls into your cluster

When you have production EMR workloads, it is recommended to enable logging, monitoring, and audit controls on your EMR cluster. This helps in debugging or troubleshooting failures, monitoring cluster usage or performance, and auditing activity for security controls.

You should integrate AWS CloudWatch for logging and monitoring, and AWS CloudTrail for audit controls. With CloudTrail audit trails, you can look for unauthorized cluster access and take the required action to harden your security implementation.

If you have integrated AWS Lake Formation on top of the Glue Data Catalog, then you should also monitor Lake Formation activity history to monitor unauthorized access on your central metastore catalog.

Blocking public access to your EMR cluster

EMR provides a great security feature, which restricts users from launching clusters with security groups that allow public access. The following screenshot shows the EMR console's **Block public access** configuration:

Figure 14.5 – The EMR console's Block public access settings

By default, all the ports are blocked, except port 22 for SSH access. You can add more ports to the exception list so that it is applicable to all clusters. You can override the port configurations through cluster security groups.

This configuration is enabled by default and is applicable to a single region of an AWS account, which means any new cluster you launch will have the same restrictions apply.

Protecting your data at rest and in transit

Data is the center of everything and protecting that is the topmost priority. When we think of protecting data, we need to consider securing it at rest and in transit.

As explained in *Chapter 7, Understanding Security in Amazon EMR*, for encryption at rest, you should encrypt your data stored in Amazon S3, HDFS, or the EMR cluster node's local disc. For data security in transit, you can configure SSL or TLS protocols.

When you consider encrypting data in transit or at rest, you should follow these best practices:

- Make sure you rotate your encryption keys regularly.

- If you do not have a specific requirement, then you can integrate **AWS Key Management Service** (**KMS**) to manage your keys, which seamlessly integrates with all AWS services, including Amazon EMR.

Having understood best practices around cluster configuration and security, next, we will explain some of the cost-optimization techniques that you can follow.

Cost-optimization techniques

There are several cost-optimization techniques that AWS offers and the primary ones are related to compute and storage resources. Let's understand some of the cost optimization techniques you can apply.

Cost savings with compute resources

When you are creating EMR on an EC2 cluster, then you have the option to choose any of the following EC2 pricing models:

- **On-Demand Instances**: This follows the pay-as-you-go model, where you pay for the EC2 instance, for the duration of time you have used it, without any commitment.

- **Savings Plans**: With **Savings Plans**, you get up to a 72% discount when you commit for a certain amount of usage ($/Hr) for 1 to 3 years. You can choose from no upfront, partial upfront, or all upfront.

- **Standard Reserved Instances**: Standard **Reserved Instances** (**RIs**) are the same as **Savings Plans** but with a variation that you must commit for specific EC2 instance type usage and also need to validate the AWS services that support RIs such as Redshift, EC2, and RDS. This offers a higher discount with less flexibility and you get a better discount with partial upfront or all upfront payment.

- **Spot Instances**: With **Spot Instances**, you can expect up to a 90% discount on your EC2 hourly cost. These are unused EC2 instances, with which you can take a risk but they also come with the risk of getting terminated.

Choosing the right type of EC2 pricing model can provide great savings. Analyze your cluster usage to identify the type of instances you should integrate and whether you have the flexibility to opt for a 1-year or 3-year commitment.

The following are some of the best practices you can follow to save costs:

- If you have persistent clusters and you are aware of the minimum number of nodes that will always be active, then look at RIs to save costs.

- If you do not have tight SLAs, then look to integrate **Spot Instances** for task nodes, which provide great discounts.

- For variable workloads, take advantage of EMR managed scaling, where you start small and scale as per your need.

- Use **On-Demand Instances** only for master and core nodes and select the appropriate instance type to optimize usage.

- Continuously monitor the usage of the CPU and memory of cluster nodes and readjust the instance type or the number of nodes as needed.

These best practices are applicable to EMR on EC2 only as with EMR on EKS, you can follow best practices for your EKS cluster. Next, we will learn what optimization techniques we can apply to the storage layer.

Cost savings with storage

There are several cost savings you can get from the storage side. The following are a few of the main ones:

- Use S3 as your persistent storage to save costs.

- Use an appropriate S3 storage class to avoid paying the cost of S3 Standard. S3 Intelligent tiering is a good option if your access patterns are not consistent.

- Define life cycle policies to move unused data to S3 Glacier for archival.

- Use columnar formats with compression, which can help with storage savings.

- If you are using cluster HDFS instead of Amazon S3, then based on the sensitivity of data or SLAs, specify different HDFS replication factors. For example, for less sensitive data that you can reproduce easily, go with replication factor 1, so that in the event of data loss, you can recover easily and can save storage costs. For highly sensitive data that has a tighter SLA, configure a replication factor of 2 or 3.

- As explained earlier, implementing partitioning provides great performance as you scan less data. It also provides great cost savings if you are planning to use Amazon Athena to query data, which has pricing based on the amount of data you are scanning.

Having understood compute- and storage-related cost savings, now let's understand what other tools AWS provides to monitor and optimize costs.

Integrating AWS Budgets and Cost Explorer

AWS Budgets and Cost Explorer are great tools to monitor costs and define thresholds to get alerted about costs going beyond the defined budget.

AWS Cost Explorer

AWS Cost Explorer is a tool that you can use to filter or group your AWS service usage costs to analyze and build visualization reporting, or you can use it to forecast your usage costs. It provides an easy-to-use web interface, where you can apply filters by AWS account, AWS service, or different date ranges and save reports for ongoing analysis. You can build some of the reports, such as monthly cost by AWS service, hourly or resource-level reports, and also a report of Savings Plans to RIs utilization costs.

AWS Cost Explorer also allows you to identify trends in usage or detect anomalies. The following is a screenshot of AWS Cost Explorer that shows its usage for the past two quarters and its forecast for the next quarter.

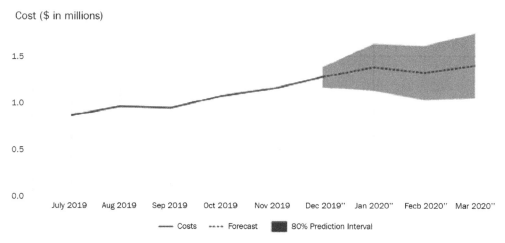

Figure 14.6 – AWS Cost Explorer report for EC2 usage

Having understood how you can benefit from AWS Cost Explorer, next let's understand how AWS Budgets can help to control your costs.

AWS Budgets

When you get started on AWS, you follow a pay-as-you-go pricing model, which means as you make progress using AWS services, your usage costs go up. As a business owner, you must have control over your costs and should have a budget defined, and if spending goes beyond that threshold, you should be notified for manual verification. This feature is offered by AWS Budgets, where you can set a custom budget and integrate notifications using Amazon **Simple Notification Service** (**SNS**) or email if actual spending or forecasted spending goes beyond the defined maximum threshold.

You can also define alarms or notifications on Savings Plans and RI usage. AWS Budgets integrates with several AWS services, including AWS Cost Explorer, to identify spending and can send notifications to your Slack channel or Amazon Chime room.

When integrating AWS Cost Explorer and AWS Budgets, you should take note of the following features and patterns:

- Use AWS Budgets to set custom budgets to track your costs and usage.

- Configure alert notifications using SNS to get notified if your usage exceeds your defined maximum budget.

- Configure notifications to be alerted if your usage of Savings Plans or RIs drops below your defined threshold.

- Use AWS Cost Explorer to continuously monitor your costs and do analysis for optimization or anomaly detection.

- AWS Budgets also natively integrates with AWS Service Catalog and that gives you the flexibility to track costs on the approved list of AWS services. AWS Service Catalog is a tool or service in AWS that allows you to define a list of services approved for usage in the organization and helps you to put restrictions on something that is not approved for usage. AWS users treat these approved services as available for use.

Having learned about AWS Cost Explorer and AWS Budgets, next we will learn about the AWS Trusted Advisor tool to understand what it offers.

AWS Trusted Advisor

AWS Trusted Advisor is one of the great tools provided by AWS. It scans your AWS service usage and provides recommendations on performance, security, fault tolerance, service limits, and cost optimization pillars.

For cost optimization recommendations, the tool looks for several factors, including the following to identify issues and potential savings:

- Low utilization of Amazon EC2 instances

- Amazon EC2 RI optimization

- Underutilized Amazon EBS volumes

- Amazon RDS database idle instances

- Amazon RDS RI optimization

- Unassociated Elastic IP addresses

The following is a screenshot of AWS Trusted Advisor, which suggests potential monthly savings with the number of checks that have passed and the number of items that need action or investigation:

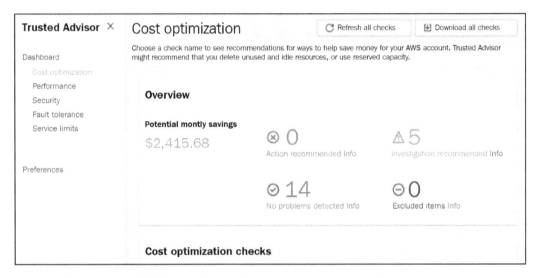

Figure 14.7 – AWS Trusted Advisor showing recommendations for cost optimization

Apart from cost optimizations, please do review recommendations suggested for performance, security, and fault tolerance to increase the stability of your AWS service implementation.

Cost allocation tags

Cost allocation tags are a unique feature offered by the AWS Billing service, which allows you to tag resources for cost calculations. It offers two types of tags – *AWS generated tags* and *user-defined tags*. AWS or AWS Marketplace **Independent Software Vendors** (**ISVs**) automatically assign tags prefixed by aws:, whereas for user-defined tags you define your custom tagging based on department, application name, cost center, or any other category.

The aws:createdBy tag is applied automatically to resources that are created after the tag has been activated. The user-defined tags created by users need to be activated first so that they appear in Cost Explorer or **Cost and Usage Reports** (**CURs**).

The cost allocation report that is generated at the end of each billing period includes both tagged and untagged resources so you can filter and order them to analyze the data based on the tags you have defined and can decide to optimize costs.

As a best practice, please make sure only authorized personnel get access to cost allocation tags in the AWS Billing console so that all the finance-related details are limited to authorized employees. Also, it's a good practice to use AWS-generated cost allocation tags.

In this section, you learned about different cost optimization techniques and available tools that can help. Next, we will cover some of the limitations that you should be aware of when you are integrating EMR for your big data analytics workloads.

Limitations of Amazon EMR and possible workarounds

Understanding best practices is very important as that helps to optimize your usage in AWS and will give you the best performance and cost optimization. Apart from best practices, it is also important to understand different limitations the service has so that you can plan for alternate workarounds.

The following are some of the limitations that you should consider while implementing big data workloads in Amazon EMR:

- **S3 throughput**: When you are writing to or reading from S3, there are a few API limits that you should be aware of. S3 has a limit of 3,500 PUT/POST/DELETE requests per second per prefix in a bucket and 5,500 GET requests per second per prefix in a bucket. These limits are per S3 prefix but there is no limit on how many prefixes you might have. So, as a workaround, you should think of having more prefixes and leverage a partition or sub-partition structure while storing data in S3. As an example, if you have 10 S3 prefixes, then you can get 55,000 read requests per second.

    Please note, these throughput limits are based on what AWS published while writing this chapter and are subject to change. Please check the Amazon S3 documentation (https://docs.aws.amazon.com/AmazonS3/latest/userguide/optimizing-performance.html) for up-to-date information.

- **EMR master failover**: EMR's cluster with multiple master nodes provide high availability but all its master nodes are on a single **Availability Zone (AZ)**. If, out of three master nodes, any two fail simultaneously or the entire AZ goes down, then your EMR cluster will go down and, in that case, it is not fault-tolerant. This is another reason to consider S3 as your persistent store and use CloudFormation to create resources so that in the event of AZ failure, you can create another cluster quickly pointing to the same S3 bucket.

- **Supported applications for multiple master nodes**: Note that EMR does not support high availability for all Hadoop or big data services it deploys and it also does not guarantee fault tolerance of the cluster services. Please refer to `https://docs.aws.amazon.com/emr/latest/ManagementGuide/emr-plan-ha-applications.html` to have up-to-date information on supported applications for multi-master nodes.

- **Cluster Monitoring**: In terms of cluster monitoring, EMR integrates with Amazon CloudWatch and provides different metrics for monitoring. But note that EMR currently does not provide any metrics specific to YARN or HDFS. So as a workaround, you can look at `MultiMasterInstanceGroupNodesRunning`, `MultiMasterInstanceGroupNodesRunningPercentage`, or `MultiMasterInstanceGroupNodesRequested` CloudWatch metrics to monitor how many master nodes are running, or on the verge of failure or replacement.

  For example, if the value of `MultiMasterInstanceGroupNodesRunningPercentage` metrics is between 0.5 and 1.0, then the cluster might have lost a master node and EMR will attempt to replace it. If `MultiMasterInstanceGroupNodesRunningPercentage` falls below 0.5, that means two master nodes are down and the cluster cannot recover and you should be ready to take manual action.

- **EMR Studio**: There are multiple considerations and limitations listed for EMR Studio. One example is EMR Studio is not supported on EMR clusters that have a security configuration attached to them with Kerberos authentication enabled. Another example is EMR Studio is not supported on an EMR cluster that is integrated with multi-master nodes or AWS Lake Formation or if the cluster integrates EC2 graviton instance types. Also note that if the EMR cluster is deployed on EKS, then it does not support SparkMagic with EMR Studio.

- **AWS Lake Formation Integration**: When you integrate AWS Lake Formation with Amazon EMR, it only supports authorization for EMR Notebooks, Apache Zeppelin, and Apache Spark through EMR Notebooks. EMR with Lake Formation does not support **Single Sign-On** (**SSO**) integration and you cannot query Glue Data Catalog tables that have partitions under a different S3 path.

- **Service Quotas and Limits**: Look for EMR service limits that can affect your production workloads. When you launch an EMR cluster on EC2, check for EC2 limits on your account. In addition, check for limits around the maximum number of concurrent active clusters you can have in your account, and in a cluster, how many maximum instances you can have per instance group.

  If you are integrating EMR SDKs or APIs, then you should be aware of the API limits it has. The two sets of limits we should be aware of are the burst limit and rate limit. The **burst limit** is the number of API calls you can make per second, whereas the **rate limit** is the cooldown period you need to have before hitting the Burst API again. For example, the `AddInstanceFleet` API call has a limit of 5 calls per second, which is the burst limit, and if you hit that limit, then you need to wait for 2 seconds before making another API call, which is called the rate limit.

These limitations are around the commonly used services or features of EMR and there will be others that you should consider while implementing your workloads. Please check the latest AWS documentation for your service before implementing it so that you can plan for alternate workarounds if possible.

Summary

Over the course of this chapter, we have learned about recommendations around choosing between transient and persistent clusters, how you can right-size your cluster with different EC2 instance types, and EC2 pricing models. We have also provided best practices around EMR cluster configurations that included cluster scaling, high availability, monitoring, tagging, catalog management, persistent storage, and security best practices.

Then, later in the chapter, we covered cost-optimization techniques that included recommendations around compute and storage, and also covered different tools AWS offers, such as AWS Cost Explorer, AWS Trusted Advisor, and cost allocation tags to monitor and control your costs with alarm notifications with AWS Budgets.

That concludes this chapter and, with it, we have reached the end of the book! Hopefully, this book has helped you to get deep knowledge of EMR's features, usage, integration with other AWS services, on-premise migration approaches, and best practices you can follow while implementing your big data analytics pipelines.

Thank you for your patience while going through this journey. If we helped you to gain knowledge, please do share your feedback and share the book with your friends and colleagues who want to get started with Amazon EMR. Thanks again, and happy learning!

Test your knowledge

Before finishing this last chapter, test your knowledge with the following questions:

1. Assume you have recently migrated your on-premise Hadoop cluster to Amazon EMR by following a lift and shift model. You have several batch and streaming workloads running on the same cluster. You have integrated your EMR cluster with AWS CloudWatch and while monitoring the cluster usage, you found not all the EC2 resources are always optimally used. What's the best architecture pattern you can follow to optimize your resource usage and costs?

2. Assume you have around five different teams who have requested to have their own persistent EMR clusters for different big data workloads. They need SSH access to the cluster master node and would like to access the web interface of Hadoop applications. How should you provide them with access while maintaining security best practices?

3. Assume you have a multi-tenant persistent EMR cluster that is deployed on EC2. It has 5 core nodes and 10 task nodes and you have enabled auto-scaling rules defined to scale the task nodes to 50 nodes as the demand arises. Which EC2 pricing model could reduce costs for you?

Further reading

Here are a few resources you can refer to for further reading:

-   Spark Performance Optimization in EMR: `https://docs.aws.amazon.com/emr/latest/ReleaseGuide/emr-spark-performance.html`

-   AWS Cost Explorer: `https://aws.amazon.com/aws-cost-management/aws-cost-explorer/`

-   AWS Trusted Advisor: `https://aws.amazon.com/premiumsupport/technology/trusted-advisor/`

-   AWS Cost Allocation Tags: `https://docs.aws.amazon.com/awsaccountbilling/latest/aboutv2/cost-alloc-tags.html`

-   EMR Studio Considerations: `https://docs.aws.amazon.com/emr/latest/ManagementGuide/emr-studio-considerations.html`

-   Considerations and limitations with AWS Lake Formation integration: `https://docs.aws.amazon.com/emr/latest/ManagementGuide/emr-lf-scope.html`

Index

A

Active Directory Federation
  Services (AD FS) 240, 245
additional custom-managed
  security groups
  specifying, for cluster 228
  working with 227
AES (Advanced Encryption Standard) 225
Airflow
  integrating, to trigger EMR
    jobs 337, 338
Amazon AppFlow 64
Amazon Athena
  about 74
  output data, querying with standard
    SQL query in 297, 298
  used, for validating output 271, 296
Amazon Athena standard SQL
  output data, querying with 272-274
Amazon Aurora
  configuring, as Hive metastore 89, 90
Amazon DynamoDB 16, 69
Amazon Elastic MapReduce (EMR)
  about 4
  advantages 9-11

Apache Ranger, setting up 248
benefits 80
comparing, with AWS Glue 21
comparing, with AWS Glue
  DataBrew 21
components 242, 243
connecting, on EC2 cluster with VPC
  endpoint interface 230, 231
connecting, on EKS cluster with VPC
  endpoint interface 232, 233
cons 26
deployment options 44
integrating, with Apache
  Ranger 247, 248
integrating, with AWS Lake
  Formation 240, 241
integrating, with AWS Glue
  Data Catalog 238, 239
Lake Formation 242
limitations 383-385
overview 8
pros 26
target users 26
use cases 26

Packt.com

Subscribe to our online digital library for full access to over 7,000 books and videos, as well as industry leading tools to help you plan your personal development and advance your career. For more information, please visit our website.

Why subscribe?

- Spend less time learning and more time coding with practical eBooks and Videos from over 4,000 industry professionals

- Improve your learning with Skill Plans built especially for you

- Get a free eBook or video every month

- Fully searchable for easy access to vital information

- Copy and paste, print, and bookmark content

Did you know that Packt offers eBook versions of every book published, with PDF and ePub files available? You can upgrade to the eBook version at packt.com and as a print book customer, you are entitled to a discount on the eBook copy. Get in touch with us at customercare@packtpub.com for more details.

At www.packt.com, you can also read a collection of free technical articles, sign up for a range of free newsletters, and receive exclusive discounts and offers on Packt books and eBooks.

Other Books You May Enjoy

If you enjoyed this book, you may be interested in these other books by Packt:

Serverless Analytics with Amazon Athena

Anthony Virtuoso, Mert Turkay Hocanin, Aaron Wishnick

ISBN: 978-1-80056-234-9

- Secure and manage the cost of querying your data
- Use Athena ML and User Defined Functions (UDFs) to add advanced features to your reports
- Write your own Athena Connector to integrate with a custom data source
- Discover your datasets on S3 using AWS Glue Crawlers
- Integrate Amazon Athena into your applications
- Setup Identity and Access Management (IAM) policies to limit access to tables and databases in Glue Data Catalog
- Add an Amazon SageMaker Notebook to your Athena queries
- Get to grips with using Athena for ETL pipelines

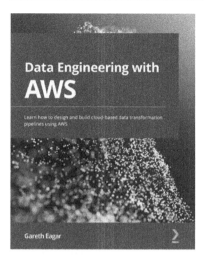

Data Engineering with AWS

Gareth Eagar

ISBN: 978-1-80056-041-3

- Understand data engineering concepts and emerging technologies
- Ingest streaming data with Amazon Kinesis Data Firehose
- Optimize, denormalize, and join datasets with AWS Glue Studio
- Use Amazon S3 events to trigger a Lambda process to transform a file
- Run complex SQL queries on data lake data using Amazon Athena
- Load data into a Redshift data warehouse and run queries
- Create a visualization of your data using Amazon QuickSight
- Extract sentiment data from a dataset using Amazon Comprehend

Packt is searching for authors like you

If you're interested in becoming an author for Packt, please visit `authors.packtpub.com` and apply today. We have worked with thousands of developers and tech professionals, just like you, to help them share their insight with the global tech community. You can make a general application, apply for a specific hot topic that we are recruiting an author for, or submit your own idea.

Share Your Thoughts

Now you've finished *Simplify Big Data Analytics with Amazon EMR*, we'd love to hear your thoughts! Scan the QR code below to go straight to the Amazon review page for this book and share your feedback or leave a review on the site that you purchased it from.

https://packt.link/r/1-801-07107-1

Your review is important to us and the tech community and will help us make sure we're delivering excellent quality content.